PERSPECTIVES IN ETHOLOGY

Volume 3

Social Behavior

CONTRIBUTORS

P.P.G. Bateson
Sub-Department of Animal Behaviour
University of Cambridge, Madingley, Cambridge, UK.

Norman Budnitz
Department of Zoology
Duke University, Durham, North Carolina

Sidney A. Gauthreaux, Jr.
Department of Zoology
Clemson University, Clemson, South Carolina

Benson E. Ginsburg
Department of Behavioral Sciences
University of Connecticut, Storrs, Connecticut

Myron A. Hofer
Albert Einstein College of Medicine
Montefiore Hospital and Medical Center, Bronx, New York

Jerome Kagan
Harvard University, Cambridge, Massachusetts

Daniel I. Rubenstein
Department of Zoology
Duke University, Durham, North Carolina

L.B. Slobodkin
Ecology and Evolution Program
State University of New York, Stony Brook, New York and
Smithsonian Institution, Washington, D.C.

Ted D. Wade
Department of Psychiatry
University of Colorado Medical Center, Denver, Colorado

Robert A. Wallace
Department of Zoology
Duke University, Durham, North Carolina

A Continuation Order Plan is available for this series. A continuation order will bring delivery of each new volume immediately upon publication. Volumes are billed only upon actual shipment. For further information please contact the publisher.

PERSPECTIVES IN ETHOLOGY

Volume 3

Social Behavior

Edited by

P. P. G. Bateson

Sub-Department of Animal Behaviour
University of Cambridge
Cambridge, England

and

Peter H. Klopfer

Department of Zoology
Duke University
Durham, North Carolina

PLENUM PRESS • NEW YORK AND LONDON

Library of Congress Cataloging in Publication Data

Bateson, Paul Patrick Gordon, 1938-
   Perspectives in ethology,

   Includes bibliographies.
   1. Animals, Habits and behavior of. I. Klopfer, Peter H., joint author. II. Title.
QL751.R188   591.5   73-79427
ISBN 0-306-36603-7 (v. 3)

© 1978 Plenum Press, New York
A Division of Plenum Publishing Corporation
227 West 17th Street, New York, N.Y. 10011

Printed in the United States of America

PREFACE

Sociobiology is the play of the season. Its success is measured by its immense popularity and perhaps by the controversy it has generated as well. Unfortunately, neither its popularity nor the resulting controversy seems likely to assure progress toward understanding sociobiological issues. The play has too many actors and, it seems, the casting has been poor; the players are unable to maintain their roles.

At center stage, of course, is E. O. Wilson and his monumental opus *Sociobiology*.[1] In the wings, and making periodic entrances, are an assortment of brilliant, committed, and aggressive adversaries. On cue, one of them steps out and decries the self-fulfilling nature of sociobiological prophesies. The arguments of the adversaries are varied. They warn that if all nonhuman primate societies tolerate aggression and man is also a primate, then aggression may come to be considered "normal" and therefore acceptable. Their dire warnings may also have real impact on policy, altering, for example, a research program intended to examine longitudinally the relation between a supernumerary chromosome and certain behavioral disorders. The rationale is that since the afflicted infants would have to be identified and the study obviously does assume that psychopathology is linked to the chromosome aberration, the attitudes of the child's parents could well contribute to abnormal behavior that might otherwise not appear.

Another group of actors in this drama introduces a second theme: man tends to misuse his tools, and thus, in the vulnerable sphere of human social relations, the tools made available must be carefully chosen. Can we ignore the manner in which the biological concept of territoriality was used by Ardrey[2] to support the political concept of private property rights? Or

[1] Wilson, E. O. (1975). *Sociobiology: The New Synthesis,* Harvard University Press, Cambridge, Mass.
[2] Ardrey, R. (1966). *The Territorial Imperative: A Personal Inquiry into the Animal Origins of Property and Nations,* Atheneum, New York.

Lorenz's (now fortunately neglected) treatise[3] urging bans on unions between persons of different "race," on the ground that such "hybridization" destroys the releasing mechanisms for the recognition of esthetic and moral ideals?

As the drama continues, yet other, socially beneficial themes could equally well emerge. And is not a clearer understanding of the nature of the human drama in itself to be desired? But how do we reconcile the disputants?

As a beginning, we should recognize that the appearance of what we term a *cause* may vary according to the vantage from which it is viewed.[4] The "cause" of a particular event differs for a historian, a biologist, and a politician, even as it may differ for biologists who operate at the molecular and the motor levels. The explicit description of these disciplinary differences, even when the focus is upon a single event, should allow for an effective isolation of political, psychological, and biological systems. Thus, facile homologizing would be prevented, and the existence of avian territoriality would not be advanced as a "cause" of capitalist political systems. The level at which disciplines interact must be superordinate to the disciplines themselves; that is, the principles governing phenomena dealt with by more than one discipline (such as territoriality) must be couched in a metalanguage or must use propositions not derived solely from the explanatory schemata of one of the disciplines.

Volume 3 of the *Perspectives in Ethology* series, we must hastily add, provides no synthesis of the sort envisaged. It does, however, provide grist for the mill from whose flour a blend can be made. We offer the grain in the belief that social harm need not follow from inquiry into the biological basis of sociality. Indeed, in the diversity of their approaches, the developmental studies of Bateson, Ginsburg, Hofer, and Kagan in and of themselves provide a caveat to the construction of homologies of man with other animals. Many of the other authors, describing animals other than man, provide equally trenchant reasons for not generalizing about one species from studies of another—even though, taken individually, none of these authors necessarily has grounds to argue this view. Finally, Slobodkin provides an important reminder that there is more than one historical dimension to biological causality. Time's arrow may not be the simple missile it appears.

In sum, the essays herein deal with important issues in sociobiology. They are offered in the belief that sociobiological issues are not only important to biologists but applicable to politics as well, providing food for

---

[3] Lorenz, K. (1940). Z. angew. Psychol. Characterkunde **59**:2–81.
[4] And note Nagel, (1961). *The Structure of Science*, Harcourt, Brace & World, New York.

thought while not fueling political polemics. We eschew discussion of these latter in the belief that the issues dealt with here (and by most biologists "doing" sociobiology) should not be considered directly relevant nor their conclusions directly applicable to political issues. That transference must await a superordinate framework.

P. P. G. Bateson
P. K. Klopfer

CONTENTS

Chapter 6

STATUS AND HIERARCHY IN NONHUMAN PRIMATE
        SOCIETIES

Ted D. Wade

Chapter 7

HIDDEN REGULATORY PROCESSES IN EARLY SOCIAL
        RELATIONSHIPS

Myron A. Hofer

Chapter 8

SOCIAL BEHAVIOR ON ISLANDS

Robert A. Wallace

Chapter 9

ON PREDATION, COMPETITION, AND THE ADVANTAGES OF GROUP LIVING

Daniel I. Rubenstein

Chapter 10

IS HISTORY A CONSEQUENCE OF EVOLUTION?

L. B. Slobodkin

homology. Hence, *cooperation* is used to refer to a social principle, with *disoperation* mediating against the development of social structure and organization within a species or any of its potentially behaviorally organizable subunits. The implications with respect to genetic mechanisms and the dynamics of selection can be quite different under the two sets of concepts, so that the distinctions are more than semantic. Whatever the explanatory concepts, the tendency to aggregate as a basic characteristic of most organisms is accepted as a given.

The aggregations themselves can take several forms. One is that in which the primitive beginnings are represented by mere coexistence for a significant period in a sufficiently tightly bounded space to permit interaction. Another is that in which the interaction itself is the central feature.

The former is represented by a variety of organisms, from simple, unicellular ones where the advantages to be derived range from conjugation through a greater tolerance for environmental variation (e.g., detoxification) as a result of the presence of the aggregate (Allee, 1931; Wilson, 1975), to infinitely more complex creatures, such as fish, many of which survive better in fish-conditioned water than in ordinary water. In each of these instances, the contribution of the aggregate to group survival is a supervening phenomenon depending upon the presence of the group.

The latter also occurs at many taxonomic levels and is, therefore, a general feature throughout the animal kingdom. Some unicellular organisms are capable of forming multicellular aggregates as well. Among these, there may even be differentiation of labor among the parts (Bonner, 1967, 1970). Schooling in fish (Nakamura, 1972), formation of migrating flocks of birds (Hochbaum, 1955), aggregations during the mating season (Scott, 1942), and changing from female to male if the previous male is removed from a group of female fish (Robertson, 1972) all represent active interactions among individuals that first form aggregations in order to functionally integrate particular activities of the group. These are fluctuating systems that are called upon in certain circumstances. Others are more permanent, as in the case of true colonial forms, from *Volvox* to coral reefs, or aggregations such as primate troops, lion prides, wolf packs, colonies of social insects, or tribes of primitive man. At this level, we have gone from aggregations to societies. The latter have both developmental cycles and typical structures. Through the evolutionary time scale, aggregates of appropriate organic molecules resulted in self-sustaining and self-replicating structures, which, in turn, aggregated and differentiated to form simple organisms. These aggregated and differentiated to form more complex organisms, and these, in their turn, form groups of various sorts in which each individual becomes a part of the structure and organization of the group. Each succeeding stage does not necessarily supplant or annihilate the former. Thus, we have viruses, protists, coelenterates, and flatworms existing today, along with termites, apes, porpoises, and men. If man represents the pinnacle of organic evolution, it is not for his individual adaptations, even including the capacity for language and the opposable thumb. It is, rather, for his interactive capacities with other members of the species, who, together, create language, art, technology, and various social structures, which have an evolution of their own. Just as man's superiority to other primates inheres in these interactive capacities, so does the termite's superiority to its primitive cockroach relatives derive from the social capabilities of the former. Man's interactive capacities may be said to be genetically based but only partly genetically determined, while those of the termite are directly built into its genes.

II. THE GENETIC ROULETTE WHEEL

Organisms vary. Particular spectra of variations keep recurring. Some variations are more frequent than others. Much variation is random and neutral from the point of view of providing any obvious advantage to the possessor. Some variations are detrimental or even lethal. Only rarely does a change occur in an already-well-adapted organism that confers some obvious advantage upon the possessor. These basic tenets of Darwinism, updated to include Mendelism, are hardly startling. Genes mutate to a restricted field of alternatives and to each with a characteristic frequency. To be preserved, they must recur until there is a reasonable pool of each mutant. To be "evaluated" as neutral, adaptive, or deleterious across the range of circumstances encountered by a species over a representative span of evolutionary time, each must occur in various combinations with all the others, since the phenotypic outcome of a gene will depend on its interaction with other genes, with developmental circumstances, and with the environmental conditions in which the species finds itself. The phenotypic value of a gene is therefore relative.

The phenotype, too, does not necessarily stand in any one-to-one relation with the genotype. Whether we are considering eye color in *Drosophila* or susceptibility to seizures in mice, the same phenotype may have multiple genetic bases (Ginsburg, 1958, 1967). Further, the genome of an individual as represented by the encoded genotype is not the same as the effective genotype or that which is translatable to phenotypic expression. For any mammalian cell, the evidence is that something on the order of 10% of its encoded genes are actually turned on. In the higher centers of the human brain, the figure is on the order of 40%, but this is not typical of other cell lines (Motulsky and Omenn, 1975). Identical genotypes of mice have variable ranges of phenotypic expression, depending upon environmental circumstances. The C57BL/10 strain is highly variable for a number of morphological and behavioral characteristics. Selection for such phenotypic variation is ineffective (Ginsburg, 1967). When identical human twins show similar variability, heritability is assumed to be low for these traits (except by the circularity of formal definition, e.g., that they are monozygotic twins). However, by analogy with the findings using mice, some of the differences in phenotype among monozygotic twins could be attributable to the fact that different environmental circumstances may activate varying components of the *encoded* genotype so that the *effective* genotypes are no longer the same. In order to know what the possibilities of any given genotype are for phenotypic expression, one would have to test it across all possible environments that it might encounter during development, thereby defining its repertoire. When one considers the gene pool of a population, therefore, its phenotypic potential depends upon the combinations and permutations of the genes themselves; upon the interactions of each encoded genome with those developmental events that will determine which aspects will come to effective expression (the "genomic repertoire," Ginsburg and Laughlin, 1971); and upon the reaction range of the effective genotype with environmental factors.

With respect to such environmental factors, the gene pool must be preadapted to circumstances not encountered in the past, such as antibiotics in the case of bacteria, and is often benefited by multiple modes of response to an adverse environmental contingency. Under conditions of high population density, for example, some mice show increased aggression and cannibalization of young, whereas others gang-rear their litters and tolerate each other well. It is probably adaptive for a population to have both of these capacities, since the first would encourage dispersal and the colonization of new areas, while the second would permit a proportion of the population to survive *in situ*.

III. THE TENDER TRAP OF ADAPTATION

Whatever the complexities by which one gets from genotype to phenotype, the prevailing notion appears to be the Darwinian one of a gradual accumulation of small adaptive changes over time, resulting in major adaptations. Ethologists have tended to anatomize and taxonomize behavior and to look for the adaptive value of each bit. Many variations are not necessarily more or less adaptive—just different. It is not necessary for all variations to be adaptive in order to persist. Some spectacularly maladaptive variations, including hemophilia and trisomy 21, persist because they continue to be genetically generated at an appreciable rate. Others, such as sickle-cell disease, confer advantages upon the heterozygote in particular environments. Many are merely neutral. Shortly after Darwinian theory became established, there was a rash of investigations designed to test some of the basic ideas. One of the better-known early investigations was that of Bumpus (1899), who measured birds killed during a storm. Most of these represented the array to be expected from the characteristics of the parent population. In other words, they constituted a random sample. Others, however, occurred in higher proportions than would be expected on a sampling basis. For the most part, these showed the greatest deviations from the mean of the population. The normative or typical bird, therefore, was the better adapted to survive.

Various studies of the stomach contents of predators have shown, contrary to expectation, that these represent a frequency sample of what is available. If more insects of a given variety happen to be present in an area, they will be found in the stomachs of birds.

We tend to overattribute adaptive value to characters without demanding proof. Darwin was aware of this and invoked other explanations, such as sexual selection or group selection, to account for variations that appeared to have no survival value *per se*. He also pointed out that with the relaxation of selection, as in the case of cave-dwelling organisms in which vision does not function, eyes may degenerate and loss of pigmentation may occur.

Returning to the point that variable modes of response to environmental conditions on the part of a species may be adaptive, selection for such variability may carry with it a tendency toward additional variability, much of which may be neutral in character.

In polygenic systems involving incremental variation, the extremes of the distribution will be the rarest phenotypes encountered. The conservative aspects of the genetic mechanisms in sexually reproducing organisms are inherently adaptive in that they preserve the array of genotypes and phenotypes characteristic of the population, which has achieved a degree of fitness. It preserves genetic variability but often buffers the range of phenotypic expression. Such examples include dominance, epistasis, penetrance, expressivity, and a variety of interactions. In the case of mammalian coat color, for example, there is generally a series of alleles in which certain combinations are necessary if pigment formation is to occur. What type of pigment (e.g., black or red) is determined by separate genes, and the intensity of each type of pigmentation is dependent on still other genes as well as on the so-called albino series (Wright, 1927; Ginsburg, 1944; Winge, 1950; Little, 1957). In the case of genes carried on separate chromosomes but affecting the phenotypic expression of the same character, at least three major types of interactions occur. Each series may make varying incremental contributions to a quantitative character. A number of interactions may produce equivalent phenotypes, as in the instances in which only the multiple recessive shows an alternative expression. There may be a threshold effect so that what would otherwise constitute a graded series if each combination of alleles came to a unique quantitative expression could constitute a phenotypic dichotomy. In all of these instances, genetic variability far exceeds phenotypic variability, enabling the species to maintain a reserve for potential variation while at the same time restricting its phenotypic variability to an adaptive range.

Any polygenic model of intelligence—assuming an inherited basis—would predict that genius is a rare event representing an extreme of a multifactorial distribution. Given its adaptive value to a species like our own, one might expect selection to have skewed the distribution towards the higher end. There are two considerations that argue against this. The first is the genetic system itself. Since the multiple heterozygote would define the mean of the population and some of the other heterozygous classes would be above the mean, phenotypic selection would tend to keep this genetic diversity, and therefore the population distribution with respect to both genotypes and phenotypes would tend to recapitulate itself. Even a high degree of assortative mating combined with genetic drift might not be sufficient to counter the equilibrium tendencies of such a system. The second consideration is that the maintenance of the system would very likely be reinforced rather than opposed by phenotypic selection, quite apart from the advantages of genetic diversity. If, as in the case of the Bumpus findings, the modal class within an array of quantitative variations is adaptive, then independently assorting alleles making incremental contribution to a particular attribute would fit this system very well. If the extreme variants should have adaptive value in particular niches, the proportion with which such variants are generated by a polygenic system could well suffice for the needs of the population. Assuming again that the biological bases of human abilities are incremental and polygenic in this way, selection would not necessarily be expected to favor an attribute such as the capacity for extreme intelligence. It is necessary for a population to have only a certain proportion of such individuals in order to optimize their contributions. Within the human species, cultural transmission will do the rest. A diversity of attributes, including the continuing ability to vary, may well be what is most adaptive. We do not have to continue to reinvent the wheel.

IV. THE PROBLEM OF HOMOGENEITY

The workings of the Hardy–Weinberg law, which forms the classical foundation for population genetics, lead to equilibrium. Under conditions of random mating, a population tends to produce the same array of genotypes and phenotypes generation after generation after generation. Different combinations of mutation and selection pressures, if constantly maintained, also lead to equilibrium in a large panmictic population (Wright, 1939). Evolutionary progress, therefore, depends upon upsetting these equilibria. Changes in environment over geologic time, isolation of one portion of the population from the rest, genetic drift, and mating patterns that provide aberrant sampling of the gene pool are among the factors that do so.

Wright (1939), in evaluating how sexually reproducing organisms could best partition their gene pools in order to promote more rapid evolution, developed a model whose essential feature is that the population does not exchange genes at random over its range but instead is subdivided into subpopulations in each of which the frequencies of many genes exist in proportions that differ from that of the population as a whole. Such subpopulations constitute the units of primary genetic exchange. They exchange some membership with adjacent subpopulations and are themselves subject to change over time. In such a system, different alleles occurring at different frequencies in the various subpopulations are "tried out" phenotypically in combination with other genes existing in high frequencies in the same subpopulations and are exposed to augmented positive or negative selection pressure. Thus the population as a whole differentiates more rapidly because of the partitioning of its gene pool. The various adaptive phenotypes arising in different parts of the population as a result of partial inbreeding and assortative mating also have a chance to be "tried out" against each other, and new combinations are formed because of the limited exchange of individuals and small groups from one subpopulation to another. Lion prides, monkey troops, wolf packs, field mice, and fish living in chains of lakes or river systems or returning to them for breeding all conform in one way or another to this type of model. The theoretical advantages from the point of view of population genetics have been pointed out by Wright (1939), and the role of social behavior in helping to establish such assortative mating systems has been discussed by Ginsburg (1968).

The behavioral aspects of populations, particularly vertebrates, that conform to the following criteria are seen as advantageous in the evolutionary sense: the promotion of partial inbreeding, the provision of the opportunity for advantageous alleles to increase their representation in the population, and the exposure of deleterious recessives to augmented selection. In the latter instance, there is also pressure for retaining heterozygosity, which promotes the retention of genetic variability while permitting phenotypic adaptation. Since genetic recombinations are necessary to produce new types or to provide the raw materials for adapting to changing environmental conditions, it becomes important for such variability to be preserved within the population, while, at the same time, it must not interfere with the general level of fitness.

V. NONRANDOM SAMPLING OF RESTRICTED GENE POOLS

Within many groups of vertebrates that are organized into functional subpopulations in the manner described, there is a further subdivision of the genetic sampling process. This subdivision occurs because in many instances, not all of the adult individuals within the pack, troop, herd, flock, etc., mate in any particular breeding season or even in a succession of breeding seasons. This irregularity further restricts the genetic sampling process. In packs of wolves, for example, we have found that over a succession of seasons, only a few of the competent females will actually produce live litters. The dominant female is among those who usually do, and in this sense, her reproductive success is related to the behavioral interactions of the group. While the picture is not the same for males, nevertheless in large packs observed for a number of breeding seasons, only a few of the males sired most of the litters as judged by the occurrence of effective ties (Rabb et al., 1967; Ginsburg, 1968).

In sage grouse (Scott, 1942), a small proportion of the males are involved in the overwhelming majority of the matings on the booming grounds. In organized flocks of domestic fowl, low-ranking individuals may not participate in the reproductive process at all (Guhl et al., 1945). While the methods for avoiding random mating differ, they are common to many organized vertebrate groups, and they depend upon behavioral adaptations. Those mentioned include the partitioning of the population into groups that may forage and hunt together or that may come together selectively for mating as in the case of sage grouse and of fish, such as salmon, that return to the spawning grounds from which they originated. Here again, the adaptation is behavioral. Within these functional aggregates, mating is not random. The social organization within each troop, pack, pride, herd, or other organized social group further restricts the number of individuals who serve as parents and therefore constitutes a funnel that restricts and determines the donors of the genes that characterize succeeding generations.

Because the genetic mechanism is so conservative, evolutionary progress over reasonable time scales depends to a large extent upon these types of behaviorally induced genetic sampling techniques. The occurrence of particular genetic combinations or of equivalent groupings of phenotypes (sometimes based on differing genetic substrata) must occur repeatedly and build up to levels of significant numerical representation if they are not to be lost as a mere matter of rarity. The persistence of particular assortative phenotypic mating patterns within the subpopulations or the augmentation of the representation of particular genetic "constituencies" by means of kin selection (see Wilson, 1975, p. 117), common paternity (Bertram, 1973, 1976; Ginsburg, 1968), and other sampling devices that also serve as multipliers would be expected to have a selective advantage for this reason. As a result (i.e., because of the social behavior of the species), not only can the effective gene frequencies become quite different among subunits of the population, but they also acquire the opportunity to build up critical numbers and to persist over a reasonable period of time.

A perusal of a number of key studies of social organization (for examples, see Wilson, 1975; Bateson and Klopfer, 1973, 1976; Bateson and Hinde, 1976) demonstrates that the social organization of a species invariably produces a partitioning of the population, particularly during the mating season, and that it further controls the systems of mating in such a way as to sample the gene pool quite differently than if mating were to occur at random and if gene flow were constrained primarily by distance. The latter circumstance, of course, has also been an important factor in this domain. Further, the organized behavior of social groups results in the repetitive sampling within each subunit of the restricted gene pool that emerges from the behavioral funnel, thereby providing another necessary condition for the effective action of natural selection.

VI. THE EVOLUTIONARY ADVANTAGES OF SOCIAL BEHAVIOR

There are many advantages accruing to an individual as a result of being a member of an organized social group. These include various devices for protection against predation, for increasing success in food getting, for providing conditions that favor the survival of the young, for access to mates, and for optimal behavioral development. From this point of view, it might be argued that selection for group behavior can be adequately explained and modeled in terms of the advantages accruing to each surviving and reproducing individual *qua* individual as a result of the behavioral adaptations of the group. In this model, there is advantageous feedback to the majority of the individuals resulting from their membership in a social group in which the whole has emerged from the parts but becomes more than the sum of the parts as a secondary elaboration resulting from individual selection. In this view, individual selection may further be seen in terms of the representation of the genes contributed by successful individuals to successive generations.

While the slow accumulation of adaptive genetic change accounts for a good many aspects of the evolutionary process, these changes themselves alter the details of the selective processes as the phenotype changes. The development of pigmented areas in association with nerve endings permits individuals to increase the efficiency of their reactions to heat and light by comparison with those who have no such specializations. From these simple beginnings, there is a long road to the vertebrate eye and to the development of binocular vision, color vision, and other sophisticated specializations for fine-tuning the environment. The specializations in the end organ are associated with specializations in the central nervous system and with emergent behavioral phenomena that are both a result of such developments and that set the stage for the selective advantages of further neural and end-organ changes. Once these visual specializations have occurred, they make many new behavioral adaptations possible, including some aspects of learning, such as the avoidance of predators and, most importantly for the subject at hand, the possibility of social adaptations involving visual cues.

Whether the capacity for detection and decoding of acoustical signals had its own initial independent evolution or not, it later became integrated with the visual and reached its peak when it became lateralized in the human brain, making the rapid decoding of speech possible and thereby carrying with it the potential for human language. The integration of these two modalities to serve the language function has resulted in the production of written language. That there are biological (genetic) parameters involved is demonstrated in familial dyslexia, which makes synthesizing the visual and the aural difficult for some individuals, who—despite normal intelligence, understanding of spoken language, and with no ascertainable defects in the visual realm—nevertheless cannot learn how to read by the usual methods. Genetic variation, then, is associated with all of these processes and their integration.

While being able to see, to hear, to judge distances, and to recognize objects improves the individual's chance for survival and provides greater control over the environment, it also makes more complex social interactions possible. This is particularly striking where the emission and recognition of social signals have a shared genetic basis, as in the case of the cricket song (Bentley and Hoy, 1974). Here, there are specific signals depending on the genotype of each form. These are preferentially responded to by the same form regardless of previous experience. F_1 hybrids between the two species studied show reciprocal differences in the courtship songs produced by the males, thereby providing evidence that X-linked genes are involved. Backcrosses between the F_1's and each of the parent species show intermediacy and conform to a polygenic model. The females, which do not themselves sing, were behaviorally tested in a choice situation. Under such conditions, F_1 hybrid females show a preferential response to songs of hybrid males of similar derivation to themselves rather than to the song of either parental species. This preference has been interpreted by Bentley and Hoy as indicating that the genes involved with the aspects of song production that they have measured and analyzed in the males may play a different role in the females, permitting them to decode, recognize, and preferentially respond to the males most closely approximating their own genotype.

In a recent symposium on "Communicative Behavior and Evolution" edited by Hahn and Simmel (1976), there is a strong emphasis on the role of communication behavior in the establishment and maintenance of social behavior in a variety of species, including our own. All highly evolved social forms depend upon such systems. In some instances, they are pheromonal; in others, such as the dance of the honey bees, the communication in relation to physical factors is more complex. On our own side of the phylogenetic tree, many researches have been made into what is inborn and what can be learned over the entire range of vertebrate species and the ways in which these communicative mechanisms are involved in various aspects of social behavior. It is not my purpose here to recapitulate these but rather to emphasize that capacities that may have initially evolved as adaptations for individuals in relation to problems of interpreting their environment can form the building blocks for the evolution of social behavior. Once communication systems become available, the function of social integration by such means becomes possible, and a whole new set of selection pressures based on social adaptations supervenes. There is then a tremendous elaboration and adaptive radiation of these behaviors with respect to social organization, social hierarchy, courtship, mating, mother–young relations, and territoriality, for example.

We have been impressed in our studies of captive wolf packs and of wolves reared in isolation (Ginsburg, 1976) with the extent to which the capacity for social behavior is a species attribute, and we have characterized the development of social behavior as a form of social genetics. The packs can develop all of the aspects of communication behavior characterizing wolves in the wild, even if left to develop without adults or with parents that have not been in contact with animals captured from the wild. In the one instance in which we have had wolf reared in isolation socialized to conspecifics that were used to receiving strangers, it, too, was able to put its communication system in place very quickly. There is a high degree of social organization in the pack structure involving highly specialized social roles, stereotyped communication, and control of group behavior. All of this organization can develop without tutors from the wild.

As one watches the social and physical development of the wolf by comparison with the dog (see Pfaffenberger *et al.*, 1976; Scott and Fuller, 1965; Mech, 1970; Rutter and Pimlott, 1968; Fox, 1975), it becomes obvious that the wolf has gone beyond the behavioral and physical development seen in dogs, which may, therefore, be considered as essentially neotenous wolves. Juvenile wolves show a close physical resemblance to some breeds of dogs, but as they grow into full maturity, they develop differences in conformation, including a broader skull, longer legs, a cow-hocked gait, and characteristic pelage that, taken together, give them quite a different appearance than they had earlier. Also at this stage, wolves begin to pay serious attention to vertical as well as horizontal surfaces and to be capable in escape paradigms of carrying out more complicated sequences of motions, involving maneuvering manipulanda on vertical surfaces, than we have ever observed in the dog. The vocal communication patterns are also different, as are many aspects of the behavior. While many of the physical gestures used are the same, they do not necessarily carry a consistent meaning in the case of the dog. In addition, the dogs become sexually mature at an earlier age, and it is possible, therefore, that their behavior is fixed at that stage and that they do not progress beyond it as the wolf does. We are now attempting to gather more detailed and critical evidence with respect to these hypotheses.

From the point of view of behavioral evolution, the pack structure of the wolf, which carries with it population control through restrictions and limitations on mating, results in a subdivision of the gene pool. Because of assortative mating within the group, mutual care of the young, and many other features, presumably including a different perceptual world and a higher order of intelligence as revealed by a problem-solving ability that appears to be beyond the dog's, behaviorally the wolf is a very different animal from his domestic counterpart. In this and other examples of organized social behavior in complex mammalian systems, it can be seen that many of the adaptations have mutuality within the group and must be thought of as involving the group as the unit of selection. Even more, the subdivision of a population into small breeding units that sample the genetic makeup of the population capriciously is here viewed as the result of selection at the species level that operates to subdivide the species in a manner that restricts the number of effective breeders; achieves partial inbreeding; samples the gene frequencies in such a way that they are different in different units of the population; provides multiplier effects including repeat matings, kin selection, common paternity, etc., so that particular combinations of genes build up to a numerical representation that would ensure adequate exposure to selective pressures. A good deal of variation is here viewed as neutral or fluctuating with respect to adaptive value and as favoring genetic systems that involve a narrower range of phenotypic expression than would be inferred from the genotypic variability. This genetic buffering permits phenotypic adaptations to occur on a variable genetic basis that provides a reserve for future genetic variation as environmental or competitive circumstances change. The evolution of social behavior (including the communication behavior that is so much a part of the animal world) from simple aggregations to complex societies is seen as primarily adaptive, because it promotes much more rapid evolution than would otherwise obtain, and would account for its relative universality. Various elaborations of social behavior can be seen as having parochial adaptive value for this or that species in other respects than partitioning and reorganizing the gene pool, but these elaborations are here viewed as derivative, whereas the major or primary function is viewed as a population genetic phenomenon as described in the Wright (1939) model, in which the possibility for diversification and evolution is speeded up by many magnitudes. Who knows whether without such partitioning devices we (*Homo sapiens*) would yet have arrived, let alone reached the stage of language and civilization that characterizes us as a species.

VII. ACKNOWLEDGMENTS

This research was supported by PHS grant MH 27591-01 and by a grant from The Grant Foundation, Inc.

VIII. REFERENCES

Allee, W. C. (1931). *Animal Aggregations: A Study in General Sociology*, The University of Chicago Press, Chicago.

Allee, W. C. (1938). *The Social Life of Animals*, W. W. Norton, New York.

Bateson, P. P. G., and Hinde, R. A. (eds.) (1976). *Growing Points in Ethology*, Cambridge University Press, Cambridge.

Bateson, P. P. G., and Klopfer, P. H. (eds.) (1973). *Perspectives in Ethology*, Vol. 1, Plenum Press, New York.

Bateson, P. P. G., and Klopfer, P. H. (eds.) (1976). *Perspectives in Ethology*, Vol. 2, Plenum Press, New York.

Bentley, D., and Hoy, R. R. (1974). The neurobiology of cricket song. *Sci. Amer.* 231:34–44.

Bertram, B. C. R. (1973). Lion population regulation. *East African Wildlife Journal* 11:215–225.

Bertram, B. C. R. (1976). Kin selection in lions and evolution. In Bateson, P. P. G., and Hinde, R. A. (eds.), *Growing Points in Ethology*, Cambridge University Press, Cambridge.

Bonner, J. T. (1967). *The Cellular Slime Molds*, 2nd ed., Princeton University Press, Princeton, N.J.

Bonner, J. T. (1970). The chemical ecology of cells in the soil. In Sandheimer, E., and Simeone, J. B. (eds.), *Chemical Ecology*.

Bumpus, H. C. (1899). The elimination of the unfit as illustrated by the introduced sparrow, *Passer domesticus*. In *Biological Lectures from The Marine Biological Laboratory, Wood's Hole, Massachusetts*, Ginn & Co., Boston.

Fox, M. W. (ed.) (1975). *The Wild Canids, Their Systematics, Behavioral Ecology and Evolution*, Van Nostrand Reinhold Co., New York.

Ginsburg, B. E. (1944). The effects of the major genes controlling coat color in the guinea pig on the dopa oxidase activity of skin extracts. *Genetics* 29:176–198.

Ginsburg, B. E. (1958). Genetics as a tool in the study of behavior. *Perspect. Biol. Med.* 1:397–424.

Ginsburg, B. E. (1967). Genetic parameters in behavioral research. In Hirsch, J. (ed.), *Behavior-Genetic Analysis*, McGraw-Hill, New York.

Ginsburg, B. E. (1968). Breeding structure and social behavior of mammals: A servomechanism for the avoidance of panmixia. In Glass, D. C. (ed.), *Genetics, Biology and Behavior Series*, Rockefeller University Press and Russell Sage Foundation, New York.

Ginsburg, B. E. (1976). Evolution of communication patterns in animals. In Hahn, M. E., and Simmel, E. C. (eds.), *Communicative Behavior and Evolution*, Academic Press, New York.

Ginsburg, B. E., and Laughlin, W. S. (1971). Race and intelligence, what do we really know? In Cancro, R. (ed.), *Intelligence, Genetic and Environmental Influences*, Grune and Stratton, New York.

Guhl, A. M., Collias, N. E., and Allee, W. C. (1945). Mating behavior and the social hierarchy in small flocks of white leghorns. *Physiol. Zool.* 18:365–390.

Hahn, M. E., and Simmel, E. C. (eds.) (1976). *Communicative Behavior and Evolution*, Academic Press, New York.

Hochbaum, H. A. (1955). *Travels and Traditions of Waterfowl*, University of Minnesota Press, Minneapolis.

Little, C. C. (1957). *The Inheritance of Coat Color in Dogs*, Cornstock Publishing Associates, Cornell University Press, Ithaca.

Mech, L. D. (1970). *The Wolf: The Ecology and Behavior of an Endangered Species*, The Natural History Press, New York.

Motulsky, A. G., and Omenn, G. S. (1975). Biochemical genetics and psychiatry. In Fieve, R. R., Rosenthal, D., and Brill, H. (eds.), *Genetic Research in Psychiatry*, The Johns Hopkins University Press, Baltimore.

Nakamura, E. L. (1972). Development and use of facilities for studying tuna behavior. In Winn, H. E., and Olla, B. L. (eds.), *Behavior of Marine Animals: Current Perspectives in Research. Vol. 2, Vertebrates*, Plenum Press, New York.

Pfaffenberger, C. J., Scott, J. P., Fuller, J. L., Ginsburg, B. E., and Bielfelt, S. W. (1976). *Guide Dogs for the Blind: Their Selection, Development and Training*, Elsevier, Amsterdam.

Rabb, G. B., Woolpy, J. H., and Ginsburg, B. E. (1967). Social relationships in a group of captive wolves. *Amer. Zool.* 7:305–311.

Robertson, D. R. (1972). Social control of sex reversal in a coral-reef fish. *Science* 177:1007–1009.

Rutter, R. J., and Pimlott, D. H. (1968). *The World of the Wolf*, J. B. Lippincott Co., Philadelphia and New York.

Schein, M. W. (ed.) (1975). *Social Hierarchy and Dominance*, Benchmark Papers in Animal Behavior, Vol. 3, Dowden, Hutchinson & Ross, Inc., Strandsburg, Penn.

Scott, J. P., and Fuller, J. L. (1965). *Genetics and the Social Behavior of the Dog*, The University of Chicago Press, Chicago.

Scott, J. W. (1942). Mating behavior of the sage grouse. *Auk* 59:477–498.

Wilson, E. O. (1975). *Sociobiology, the New Synthesis*, Belknap Press of Harvard University Press, Cambridge, Mass.

Winge, O. (1950). *Inheritance in Dogs with Special Reference to Hunting Breeds*, Roberts, C. (translator), Cornstock Publishing Co., New York.

Wright, S. (1927). The effects in combination of the major color factors of the guinea pig. *Genetics* 12:530–569.

Wright, S. (1939). *Statistical Genetics in Relation to Evolution*, Hermann, Paris.

Chapter 2

THE ECOLOGICAL SIGNIFICANCE OF BEHAVIORAL DOMINANCE

Sidney A. Gauthreaux, Jr.

Department of Zoology
Clemson University
Clemson, South Carolina 29631

I. INTRODUCTION

In a discussion of the evolution of gene flow, Wilson (1975, p. 103) pointed out that a tendency for different sexes and age groups to migrate differentially can exert a profound influence on social structure. In this paper, I suggest that the reverse is equally true: a basic social structure—namely, the dominance hierarchy—can strongly influence the differential dispersal and migration of different sexes and age groups and thereby serve an important function in population dynamics.

Although dominance hierarchies have been routinely thought of in terms of a localized group of individuals within a population, in this paper I present evidence that the characteristics of a dominance rank may also be applied at the level of the entire population. Just as individuals within a localized group have different ranks in a dominance hierarchy depending on their aggressiveness, age, sex, size, familiarity with the area, family relationship, health, and social skills, among other factors (Collias, 1944; Allee *et al.,* 1949; Thompson, 1960; Brown, 1963, 1975; Wilson, 1975; Bernstein, 1976), the individuals that comprise a species may have different ranks depending on the same criteria. At the population level, age and sex are unquestionably important determinants of dominance status, but a given individual's status in the population may be raised or lowered in relation to its class rank depending on additional factors (e.g., aggressiveness or competitiveness, familiarity with the area).

Application of the concept of dominance rank at the level of the entire population permits the construction of a graphical model that interrelates dominance rank, the quantity of a potentially limiting resource, dispersal, and migration. Moreover, invoking dominance rank as an underlying regulatory factor in the population dynamics of a species can explain differential habitat utilization or geographic distribution during the nonbreeding season, differential mortality during the nonbreeding and breeding seasons, the orderly return of individuals to the breeding grounds or breeding habitat, the establishment and quality of the individual's territory, mating success, and ultimately the individual's reproductive success. All of these events are of course interrelated, and as will be shown, dominance behavior can regulate each directly or indirectly. It is important to stress that the annual cycle of a species is a closed loop, that the events regulated by dominance rank during the nonbreeding season feed back on the breeding season, and that in turn the events regulated by dominance rank during the breeding season have effects during the nonbreeding season. The idea as outlined applies more readily to animals that survive for more than one year in a seasonal environment and have several reproductive periods (hence, experience can be acquired and social rank in the hierarchy can change), but the hypothesis can also be applied to animals that have a relatively short life span and can reproduce only once in a more constant environment.

II. HISTORICAL CONSIDERATIONS

The formulation of the concept of social dominance began with the work of Pierre Huber (1802), who first recognized dominance orders during his studies of bumblebees. Further work showed that the dominance relationship among bumblebees was orderly and predictable (Hoffer, 1882). This work on social dominance by two entomologists essentially went unnoticed, and it was primarily the work of Thorleif Schjelderup-Ebbe on dominance hierarchies in flocks of chickens, *Gallus domesticus* (1922a,b), and in flocks of wild and tame ducks, *Anas bosehas* (1923), that attracted the attention of behaviorists and ecologists to the concept. Schjelderup-Ebbe's studies demonstrated a rather highly structured dominance order or hierarchy in which each individual responded to every other member in either a dominant or a subordinate manner. The organization was called a peck order, because the dominants tended to peck subordinates. This, however, happened only for a few days; once the peck order of the flock was established, the pecking was greatly reduced. Males tended to dominate females, and older birds dominated younger ones. Thus, the concept of dominance demonstrated that one animal had preferential rights over another. The dominant animal had first choice of desired limited resources, and the subordinate "accepted" the arrangement. When the sought-after resource was plentiful, both individuals existed together and shared the resource; when the resource was limited, the subordinate did without.

Following the classic work of Schjelderup-Ebbe on chickens and ducks, there was considerable effort to confirm and to extend to other species the concept of social dominance. Some workers in the same decade continued to elaborate on the hierarchial arrangement in chickens (Katz and Toll, 1923; Fischel, 1927; Kroh, 1927). Schjelderup-Ebbe (1931) himself extended his observations to include many different species of birds both in nature and in captivity, and he concluded that social dominance was one of the fundamental principles of biology.

In the 1930s, there began a surge of publications on social dominance. Konrad Lorenz (1931, 1935) discovered that individual birds in jackdaw (*Corvus monedula*) colonies also had a strict rank order. He found that the highest ranking birds were peaceful to very low-ranking birds but that high-ranking birds were more aggressive toward birds ranked immediately below them. With regard to mating, males chose females of a rank lower than their own. A low-ranking jackdaw female at once advanced in the hierarchy when she mated with a higher ranking male, and she changed her behavior accordingly.

A largely descriptive paper on social dominance in chickens (*Gallus*) and pigeons (*Columba*) by Masure and Allee (1934) served as the basis of several subsequent papers by W. C. Allee and his colleagues at the University of Chicago in which the relationship between hormones and social rank (e.g., Allee and Collias, 1938, 1940; Allee et al., 1939; Allee et al., 1940) as well as the role of prior experience and dominance rank (e.g., Ginsburg and Allee, 1942; Guhl and Allee, 1944; Guhl et al., 1945) was emphasized. These papers concluded that androgens and rank position were highly correlated. In actual matings, high-status female chickens tended to be courted and mated less frequently than low-status females with less androgen. High-status males mated more frequently and produced more offspring than did low-status males, and the high-ranking males tended to "psychologically castrate" low-status males.

During the middle 1930s, Carl Murchison and his colleagues at Clark University published a series of papers that represents the earliest attempts to quantify and to statistically analyze the measurable and identifiable characteristics correlated with dominance (Murchison, 1935a,b,c; Murchison et al., 1935). Murchison eventually derived a general equation that accounted for the contribution of the predictor variables to the rank of the individual animal (Murchison, 1935d, 1936). Unfortunately, this attempt to quantify and model the phenomenon of dominance rank did not stimulate additional work until the quantitative work of N. Collias, H. G. Landau, N. Rashevsky, and I. Chase.

In an attempt to gain insight into the factors that decide the outcome of initial encounters Collias (1943) used statistical analysis to evaluate 200 pair contacts of white leghorn hens in a mutual pen. Controlled factors included sex, territorial familiarity, and social facilitation. The influences of the more important factors were determined by the method of path coefficients, a method fundamentally similar to multiple regression. The important factors were male-hormone output and thyroxin secretion. Social rank in the home flock had much less influence, and weight was of only small importance. The multiple correlation coefficient of success with the four factors analyzed was 0.75. In a series of papers dealing with the mathematical analysis of social behavior, Landau (1951a,b, 1965, 1968) showed that the orderly hierarchies found in many animal groups cannot be easily explained even with a full knowledge of the determinants and their correlations with fighting ability. Landau further demonstrated that as the size of the group increases the strength of the hierarchy declines sharply. He concluded that social factors, or psychological factors such as the previous history of dominance, which were not included in his treatments, were probably of greater importance in explaining the observed hierarchies in flocks of domestic hens (Landau, 1951a). The effectiveness of social bias in establishing hierarchies was considered much greater in small societies than in large ones (Landau, 1951b). Landau (1965) also considered the development of hierarchies when new members are added successively. He suggested that if the new member dominates all the older ones below a certain rank and is dominated by all those above this rank, the hierarchy will persist if it is the initial structure. As the size increases, the structure will tend to hierarchy if it is not the initial structure. In a later paper Landau (1968) examined three models of possible development of a society with a dominance relationship and concluded once again that social factors as well as inherent characteristics need to be introduced to account for near-hierarchical structures. Chase (1974) also examined two major models of dominance hierarchies (tournament model and correlation model) and concluded that severe mathematical conditions are required to generate strong hierarchies and that existing data on hierarchies do not meet the conditions. He argued that pairwise competitions, rankings of individuals, and round robins are all artificial constructs and different in form from the actual process of hierarchy formation in nature, and he suggested that future research should examine more closely the natural social situation in which hierarchies develop. Rashevsky (1959) has discussed some general mathematical considerations of social organizations, including dominance hierarchies, in a book devoted to the subject.

Numerous studies on various animal species since the middle 1930s have thoroughly documented the ubiquity of the phenomenon of social dominance in the animal kingdom. Interestingly, during this period dominance was rediscovered in social insects (*Polistes*) by G. Heldmann (1936a,b). Excellent reviews of one or more aspects of social dominance can be found in Allee (1939, 1942), Noble (1939), Collias (1944), Guhl (1953), Brown (1963, 1975), Klopfer and Klopfer (1973), Schein (1975), Wilson (1975), Bernstein (1976), and Ralls (1976). Although most studies on dominance hierarchies have been conducted in the laboratory, Collias (1944), Brown (1963, 1975), Thompson (1960), and Wilson (1975) have reviewed a number of field studies of dominance relationships among birds and other animals in natural populations.

Table I gives information on the array of species in which dominance has been shown. The list is not intended to be exhaustive but is nonetheless quite representative. One thing should be emphasized: dominance hierarchies appear to be distributed through the animal kingdom in a highly irregular fashion. The distribution undoubtedly is in part due to the lack of work on certain groups and hence the state of our knowledge, but in some cases social complexities may mask the traditional manifestation of dominance relationships (e.g., in many primates, Bernstein, 1976). The occurrence of dominance hierarchies, like other forms of social organization within a species, is influenced by the recent evolutionary history of the species and the ecological setting within which it occurs. It is apparent that the ecological and evolutionary importance of social dominance warrants greater emphasis, and in the following sections, these aspects of behavioral dominance are stressed.

III. BEHAVIORAL DOMINANCE AS A NATURAL REGULATORY SYSTEM

There is general agreement that the dominance rank (dominance hierarchy or social hierarchy) is the set of sustained aggressive–submissive relations among a group of animals that results most commonly in multiple ranks in a quasi-linear sequence occasionally containing triangular or other circular elements (Wilson, 1975). The hierarchy is formed by initial aggressive encounters, albeit sometimes of a subtle and indirect nature, and once the individuals assume their ranks in the group, there is little addi-

Table I. A Representative Listing of Species Showing Social Dominance

---

MOLLUSCA
  Octopoda:
    Octopodidae, octopus—Yarnall (1969)
ARACHNIDA
  Araneida:
    Thomisidae, crab spider—Braun (1958)
    Linyphiidae, sheet-web spider—Rovner (1968)
CRUSTACEA
  Decapoda:
    Homaridae, lobster—Douglis (1946)
    Palinuridae, spiny lobster—Fielder (1965)
    Astacidae, crayfish—Bovbjerg (1953, 1956); Lowe (1956)
    Paguridae, hermit crab—Allee and Douglis (1945); Hazlett (1966, 1970)
INSECTA
  Orthoptera:
    Blattidae, cockroaches—Ewing and Ewing (1973)
    Tettigonidae, katydids—Morris (1971)
    Gryllidae, crickets—Alexander (1961); Alexander and Thomas (1959)
  Lepidoptera:
    Nymphalidae, butterflies—Crane (1957)
  Coleoptera:
    Scarabaeidae, beetles—Beebe (1947)
  Hymenoptera:
    Formicidae, ants—Brian (1952); Lange (1967)
    Vespidae, wasps—Heldmann (1936a,b); Pardi (1940, 1948); Deleurance (1948, 1952);
      Gervet (1956, 1962); Scheven (1958); Morimoto (1961a,b); Yoshikawa (1963);
      Montagner (1966); Eberhard (1969); Wilson (1971)
    Apidae, bees—Huber (1802); Hoffer (1882); Sakagami (1954); Free (1955, 1961); Free and
      Butler (1959); Sakagami et al. (1965)
PISCES
  Squaliformes:
    Carcharhinidae, sharks—Allee and Dickinson (1954)
  Salmoniformes:
    Salmonidae, trout—Jenkins (1969)
  Cypriniformes:
    Gymnotidae, electric fish—Westby (1974)
  Atheriniformes:
    Poeciliidae, livebearers—Braddock (1945, 1949); Baird (1968); Thines and Heuts (1968);
      McKay (1971); Farr and Hernnkind (1974); Gorlick (1976)
  Perciformes:
    Centrarchidae, sunfishes—Greenberg (1946); Erickson (1967); McDonald et al. (1968)
    Cichlidae, cichlids—Figler et al. (1976)
    Pomacentridae, damselfishes—Myrberg (1972); Moyer (1976)
    Blenniidae, combtooth blennies—Gibson (1968)
AMPHIBIA
  Anura:
    Pipidae, clawed frog—Haubrich (1961)
    Ranidae, frogs—Boice and Witter (1969); Boice et al. (1974)
    Bufonidae, toads—Alexander (1964); Boice et al. (1974)

---

continued

Table I (*Continued*)

REPTILIA
Testudinata:
Chelydridae, snapping turtles—Froese and Burghardt (1974)
Testudinidae, land tortoises—MacFarland (1972); Boice *et al.* (1974)
Squamata:
Iguanidae, lizards—Evans (1936)

AVES
Sphenisciformes:
Spheniscidae, penguins—Sladen (1958); Penny (1968); Spurr (1974)
Pelecaniformes:
Sulidae, boobies—Simmons (1970)
Ciconiiformes:
Ardeidae, night herons—Noble *et al.* (1938)
Anseriformes:
Anatidae, ducks and geese—Jenkins (1944); Boyd (1953); Collias and Jahn (1959); Raveling (1970)
Falconiformes:
Accipitridae, vultures—Valverde (1959)
Galliformes:
Tetraonidae, grouse—Johnsgard (1967)
Phasianidae, pheasants—Collias and Taber (1951); Banks (1956); Collias and Collias (1967); Hughes *et al.* (1974)
Meleagrididae, turkeys—Watts and Stokes (1971)
Gruiformes:
Gruidae, cranes—Miller and Stephen (1966)
Charadriiformes:
Laridae, gulls—Noble and Wurm (1943); Coulson (1968)
Columbiformes:
Columbidae, doves and pigeons—Bennett (1939, 1940); Murton *et al.* (1966); Goforth and Baskett (1971)
Psittaciformes:
Psittacidae, parakeets—Masure and Allee (1934)
Cuculiformes:
Cuculidae, ani—Davis (1942)
Passeriformes:
Formicariidae, antbirds—Willis (1967, 1968)
Paridae, titmice—Colquhoun (1942); Odum (1942); Brian (1949); Kluyver (1957); Dixon (1963, 1965); Hartzler (1970); Minock (1971); Glase (1973); Smith (1976)
Zosteropidae, white-eyes—Kikkawa (1961); Williams *et al.* (1972)
Emberizidae, sparrows—Tinbergen (1939); Sabine (1949, 1959); Dunham (1966)
Fringillidae, finches—Shoemaker (1939); Tordoff (1954); Hinde (1955, 1956); Marler (1956); Nicolai (1956); Thompson (1960)
Ploceidae, weaverbirds—Ward (1965); Watson (1970); Dunbar and Crook (1975)
Sturnidae, starlings—Davis (1959)
Corvidae, crows and jays—Lorenz (1931, 1935); Goodwin (1951); Lockie (1956); Brown (1963)

MAMMALIA
Marsupialia:
Macropodidae, kangaroos and wallabies—Russell (1970); LaFollette (1971); Grant (1973)

continued

Table I (*Continued*)

Insectivora:
Tupaiidae, tree shrews—Sorenson (1974)
Primates:
General—Maslow (1935, 1936, 1940); Gartlan (1968); Chance and Jolly (1970); Jolly (1972); Bernstein (1970, 1976)
Lemuridae, lemurs—Jolly (1966); Sussman and Richard (1974)
Indridae, Indrid lemurs—Sussman and Richard (1974)
Lorisidae, galago—Roberts (1971); Drews (1973)
Cebidae, New World monkeys—Maslow (1936, 1940); Carpenter (1942a,b); Eisenberg and Kuehn (1966); Baldwin (1971)
Cercopithecidae, Old World monkeys—Maslow (1936, 1940); Carpenter (1942a,b); Altmann (1962); Rowell (1963, 1971); Sade (1967); Southwick and Siddiqi (1967); Poirier (1970); Alexander and Hughes (1971); Bartlett and Meier (1971); Chalmers and Rowell (1971); DeVore (1971); Vessey (1971); Jolly (1972); Dunbar and Dunbar (1976)
Pongidae, great apes and gibbons—Maslow (1936, 1940); Carpenter (1942a,b)
Lagomorpha:
Leporidae, rabbits and hares—Mykytowycz (1962); Myers *et al.* (1971)

Rodentia:
Sciuridae, squirrels—Pack *et al.* (1967); Farentinos (1972); Michener (1973); O'Shea (1976)
Heteromyidae, kangaroo rats—Blaustein and Risser (1976)
Cricetidae, New World rats, mice, and hamsters—Eisenberg (1963); Lawlor (1963); Sheppe (1966); Archer (1970); Grant (1970); Sadleir (1970); Christian (1971); Drickamer and Vandenbergh (1973); Drickamer *et al.* (1973); Kinsey (1976); Rowley and Christian (1976)
Muridae, Old World rats and mice—Uhrich (1938); Davis (1958); Calhoun (1962); Barnett (1963, 1967); Baenninger (1966, 1970); Reimer and Petras (1967); Becker and Flaherty (1968); Archer (1970); DeFries and McClearn (1970); Boreman and Price (1972); Oakeshott (1974); Syme *et al.* (1974); Poole and Morgan (1976); Price *et al.* (1976)
Carnivora:
Canidae, wolves and dogs—James (1939); Schenkel (1947); Antonov (1976)
Hyaenidae, hyenas—Kruuk (1972)
Felidae, cats—Leyhausen (1956, 1971)
Otariidae, sea lions—Peterson and Bartholomew (1967)
Phocidae, elephant seals—LeBoeuf and Peterson (1969); LeBoeuf (1972)

Perissodactyla:
Equidae, horses—Grzimek (1949)
Artiodactyla:
Suidae, swine—McBride (1963); Meese and Ewbank (1973); Ewbank *et al.* (1974)
Camelidae, llamas—Pilters (1954)
Cervidae, deer and elk—Darling (1937); Altmann (1952); Schaller (1967); Vos *et al.* (1967)
Bovidae, bison, sheep, goats, and cattle—Scott (1945, 1946); Schein and Fohrman (1955); McHugh (1958); Schloeth (1958); Schaller (1967); Fraser (1968); Geist (1971); Klopfer and Klopfer (1973); Estes (1974)

tional aggression unless a perturbation occurs. In absolute dominance hierarchies, the rank order is the same wherever the group goes (location independent) and whatever the circumstances, but in relative dominance hierarchies, the rank order is dependent on place (location dependent) and circumstances (Brown, 1963, 1975; Wilson, 1975). Both Wilson (1975) and Brown (1975) regard dominance behavior as the analogue of territorial behavior, differing in that the members of an aggressively organized group of animals coexist within one area with resources shared and not divided. If subordinates are not excluded, they remain in a subordinate status but typically at the cost of restricted access to the essentials for reproduction and survival.

Recently, considerable attention has been directed to the clarification of the numerous and confusing behavioral definitions of dominance and to the means of measuring it (Chase, 1974; Richards, 1974; Rowell, 1974; Syme, 1974; Bernstein, 1976; Drews and Dickey, 1977). The basis for establishing dominance hierarchies has been traditionally agonistic behavior, either overt or covert. In laboratory studies, staged encounters between pairs of animals have often been used to generate a matrix for all the individuals within a group, the total number of wins and losses determining the rank of the individual in the group. In field studies, the hierarchy has been determined by a scoring of the winners and losers in interactions of marked, free-living individuals at a feeding tray or bait. The types of behavior used by observers to judge dominance are as diverse as the species observed (Richards, 1974). Given the questionable utility of some of the competitive measures employed in the laboratory as well as the ultimate intent of being able to use the dominance concept in naturalistic settings, Jensen (1970) and Drews and Dickey (1977) have argued for a more observationally based measure of dominance with some descriptive breadth as the foundation of an empirically based dominance concept. Such a measure will certainly help, but consideration of the evolutionary and ecological implications of behavioral dominance is equally important. In general, dominance implies priority of access acquired by past or present competitive or cooperative abilities in nature rather than by chance or by some factor unrelated to dominance. Thus, measures of dominance must (1) indicate priority unambiguously and (2) indicate that the observed priority is caused, at least indirectly, by the dominance behavior of the participants rather than by some incidental factor.

The concept of dominance can be applied to small groups of individuals, to large groups of individuals, to the entire population, and even to two or more competitive species. Furthermore, the determinants of dominance may vary depending on the type and size of the group of individuals. In Table II, I have attempted to classify the characteristics that contribute

Table II. Determinants of Dominance Organization

INTRASPECIFIC	Individual Characteristics—only in small groups; individual characteristics used for recognition; aggressive encounters initially determine rank, and status is maintained by individual recognition.
	Class Characteristics—groups may be small to large; supraindividual characteristics are important in determining and maintaining status (e.g., sex, age, size, display, etc.); aggressive encounters may occur but are not necessary.
	Group Characteristics—generally a social unit in itself (family, caste, troop, etc.); a "super organism" with divisions of labor; social skills and group size are important determinants of dominance; intergroup aggressive encounters could establish dominance of a group within a given area.
INTERSPECIFIC	Species Characteristics—entire species; species-specific characteristics are important; interspecific aggressive encounters may or may not be necessary to determine dominance gradient.

to intraspecific and interspecific dominance organizations. The classification is based on the belief that social dominance is a behavioral means of reducing intraspecific and interspecific competition and costly interactions. Social dominance is analogous to the evolutionary process of adaptive radiation in that it permits a group of individuals (or species) to coexist in an orderly fashion with minimal deleterious interactions while permitting the most efficient utilization of resources. Unlike adaptive radiation, the individuals (or species) are not locked into an exploitation strategy and can change their strategy depending on competitive circumstance. In this regard, social dominance is a plastic strategy much the same as learning. I believe behavioral dominance is a very basic organization in nature with profound ecological implications and is the result of competitive characteristics of the individual, the class, the group, and even the species.

The classification in Table II suggests that within a rather small group of individuals a dominance organization can be established initially by agonistic interaction and maintained by individual recognition. Although individual characteristics can be the basis of a dominance hierarchy, Wood-Gush (1971) and Hughes et al. (1974) have pointed out that individual recognition is probably of lesser importance in large flocks of chickens, where dominance relationships depend more on individuals' identifying the status of their flockmates by their demeanor, structural features, and displays rather than by recognizing them as individuals. Such characteristics as sex, age, and size I have classified as class characteristics. The majority of intraspecific dominance hierarchies are determined by individual and class factors. When more elaborate social organizations exist (e.g., family, caste, troop), the attributes of the social unit can also determine dominance rank. In primate societies, social skills are an important determinant of dominance (Bernstein, 1976), and in conspecific competition, large troops generally dominate small troops (Altmann, 1962; Wilson, 1968). Thus, in cases concerning social groups, the dominance hierarchy is probably determined by a combination of characteristics of the individual, the class, and the group. Although there may be a dominance organization within the group, the intergroup hierarchy is equally important as a regulatory system. Whenever it is possible to speak of an interspecific dominance hierarchy (McHugh, 1958; MacMillan, 1964; Willis, 1967, 1968; Moore and McKay, 1971; Barlow, 1974; Morse, 1974), species-specific characteristics are undoubtedly the determinants of the hierarchy. These may in some cases be identical to certain class characteristics that function in the development of intraspecific dominance hierarchies.

The importance of hierarchies of social rank in the regulation of population density, dispersal, and ultimately the evolution of mammals has been stressed by Christian (1970). In this paper, I suggest as did Christian (1970) that the concept of dominance hierarchies has utility at the level of the species. The subsequent sections of this paper emphasize the role of dominance rank at the population level in terms of dispersal, irruptions, short- and long-distance migration, and homing. This is the first attempt to place these types of behavior in their proper ecological perspective, and as a result, their importance to the remainder of the species' annual cycle can be discussed.

IV. THE DOMINANCE–DISPERSAL MODEL

If behavioral dominance or class rank at the level of the population is indeed an important factor in influencing differential dispersal of adults, young, males, and females, then how does the hierarchy relate to dispersal? In Fig. 1, I have constructed a graphical model that helps to explain the interrelations of dominance rank, dispersal, and a limiting resource. The model is discussed first, followed by a review of supportive data.

Of fundamental importance to the model in Fig. 1 is that the ordinate is a resource (e.g., food) that is potentially limiting and can generate intense competition. Priority of access to this resource is the major criterion for dominance status in the scheme. I will begin the explanation of the model by considering the resource to be food and the species to have just completed its breeding season in the temperate zone. If the quantity of food in the breeding habitat is high, all members of the dominance hierarchy can remain in the habitat and share the food resources in an orderly manner

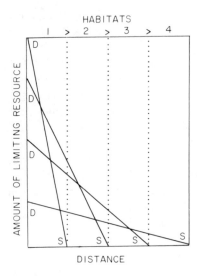

Fig. 1. Model of dominance-influenced dispersal. The lower abcissa represents geographical distance; the upper abcissa represents habitats of decreasing quality. The ordinate is the amount of limiting resource on the breeding grounds. The different slopes represent dominance hierarchies, with the most dominant (D) at the top of the slope and the most subordinate (S) at the bottom.

according to rank with minimal competitive interaction or aggression. Adults and young of both sexes can coexist in the same habitat. It should be stressed that even though all individuals are occupying the same habitat in the breeding range, there is the tendency for the most subordinate individuals to be found on the periphery of the area and to have moved the greatest distance (Ward, 1965; Murton *et al.,* 1966; Dunbar and Crook, 1975).

As food supplies decrease with increasing time, the most subordinate animals cannot acquire sufficient food, and they become hyperactive and emigrate. The dispersal may carry these subordinate individuals to areas where they can obtain food resources in the same habitat type but in a different geographical area (lower abcissa), or the subordinates may shift to a lesser quality habitat without individuals of higher dominance rank and have free access to a somewhat different or reduced food supply (upper abcissa). Thus, during a period of reduced food, the dominance rank of an individual can be expressed in terms of the distance it has moved from its place of birth or in terms of the quality of the habitat it occupies, or both. Dominants are close to their place of birth in prime habitat, while subordinates, forced to emigrate, occupy areas farther away from the breeding areas in poorer quality habitats.

The most dominant individuals remain in the breeding habitat as long as sufficient food for survival is present. However, when food in the prime habitat is suddenly, temporarily eliminated (e.g., by a severe winter snowstorm that covers food), the dominants temporarily move to areas that permit access to food and return to the prime habitat as soon as conditions improve. I view such responses as incipient true migratory movements, and with regard to the model (Fig. 1), the ordinate is shifted to the right along the abcissa (Fig. 2). When the time span with no food supply is of seasonal duration (e.g., winter) and occurs every year, the pattern of movement from and to the breeding area is called migration.

In the winter range of the species, the dominance rank can still be expressed in terms of distance to the breeding site or of habitat quality, or both. In spring, when individuals return to the prime habitat to breed, the dominants have priority of access to the best breeding sites. Some of the most dominant individuals may have managed to maintain the best breeding sites during the nonbreeding season. This is particularly true if the best site during the breeding season proved to be the best site during the nonbreeding season, but not necessarily. The order of arrival of individuals on the breeding grounds follows the dominance gradient, with the most dominant arriving first (if it was not already there), followed by successively more subordinate individuals. Some of the most subordinate individuals may not attempt to return to the breeding grounds because of the poor energetic condition they are in at the end of the nonbreeding season. The subordinates that arrive late are less likely to have a successful reproductive season because of lack of females, poor nesting habitat, heavier predation on young produced late in the nesting season, and other factors.

The model can also be applied to irruptions, partial migrations, and short-distance and long-distance migrations, and the following predictions

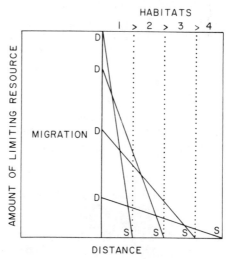

Fig. 2. Model of dominance-influenced migration. No food is available on the breeding grounds during the nonbreeding season, so that the dominance gradients are shifted geographically to a point where food becomes available. The ordinate in this figure represents the quantity of limiting resource in the prime non-breeding or wintering habitat. For additional explanation see Fig. 1.

are possible:

1. The individuals that move the least distance and remain in the prime habitat are the dominants and possess characteristics that confer dominant status with regard to size, sex, age, and familiarity with the area, and the individuals that move the greatest distance (or into the least favorable habitat) are the subordinates and possess characteristics that are correlated with subordinate status with regard to size, sex, age, and familiarity with the area.

2. If the nonbreeding range of the species is geographically disjunct from the breeding range, the dominants acquire areas closest to the breeding grounds or in prime habitats with respect to food in the nonbreeding range. The subordinates are forced to areas farther away from the breeding grounds or into habitats of less quality with respect to food in the nonbreeding range.

3. Because of the differential distance between breeding grounds and nonbreeding grounds for dominants and subordinates, or because of the differential food availability for dominants and subordinates on the nonbreeding grounds, dominants return to the breeding grounds before subordinates.

4. Because dominants arrive on the breeding grounds before subordinates, they have priority of access to the best breeding sites and are most capable of defending these sites.

5. If the nonbreeding range of the species is geographically disjunct from the breeding range and dominants acquire the areas in the nonbreeding range closest to the breeding grounds, then during postbreeding movement the subordinates precede the dominants.

6. If the nonbreeding range of the species is geographically disjunct from the breeding range and dominants acquire areas in prime habitats with respect to food in the nonbreeding range, then during postbreeding movement the dominants arrive on the nonbreeding grounds before the subordinates.

7. The dominants are likely to show better homing tendencies to successful breeding sites than subordinates.

8. If dominants show well-developed homing to nonbreeding areas, then the species segregates by habitat rather than by geographical distance during the nonbreeding season.

V. SUPPORTIVE DATA AND DISCUSSION

Considerable data exist to support the idea that dominance rank is an important factor in regulating dispersal; irruptions; and partial, short-, and long-distance migrations of many animal species. The characteristics of these types of behavior are explainable in terms of the characteristics of dominance rank. The following portion of this paper clearly documents the similarities.

In the dispersal of individuals from a population, an interesting and important peculiarity is often evident: a leptokurtic distribution of individuals after some time period following the beginning of dispersal (e.g., Dobzhansky and Wright, 1943, *Drosophila pseudoobscura*; Dice and Howard, 1951, prairie deermice, *Peromyscus maniculatus*; Johnston, 1956, song sparrows, *Melospiza melodia*; Aikman and Hewitt, 1972, grasshoppers, *Myrmeleotettix maculatus*; Johnston and Heed, 1975, *Drosophila nigrospiracula*). These studies draw particular attention to the leptokurtosis in the dispersal pattern, but earlier studies by Nice (1937) on song sparrows and Gross (1940) on herring gulls, *Larus argentatus*, show basically the same pattern. The leptokurtosis is usually attributed to two types of individuals in the population, those tending to disperse fast (highly mobile animals) and those tending to disperse much slower (more sedentary animals). Evidence for genetic differences between dispersing and resident individuals within a population has been presented for insects (e.g., Uvarov, 1961; Wellington, 1964; Dingle, 1965, 1966, 1968; Shaw, 1967, 1968; Den Boer, 1968; Narise, 1968; Gilbert and Singer, 1973; Waloff, 1973) and mammals (Thiessen, 1964; Myers and Krebs, 1971; Krebs *et al.*, 1973). In addition, several models (Parsons, 1963; Maynard Smith, 1964, 1974; Levins, 1970a,b; Van Valen, 1971; Roff, 1975) have elaborated the role of a genetic polymorphism for dispersal and migration traits within a population in terms of migrant selection.

Although a polymorphism can ultimately explain both sedentary and motile individuals within a population, the differential rate of dispersal within a population can also be influenced more directly by social or behavioral dominance. Behavioral dominance can play an indirect role in resource partitioning among mobile animals by directly influencing their patterns of movement during dispersal, irruptions, and short- and long-distance migrations. Moreover, the status or rank of the individual can change in response to environmental change or a change in age.

In his review and reevaluation of territorial behavior and population regulation in birds, Brown (1969) discussed the importance of breeding and nonbreeding seasons in treatments of spacing systems in relation to important resources. He suggested that the critical periods for postbreeding losses (after postjuvenal molt) occur in two phases, as demonstrated by the data from Gibb (1960, 1962). In Phase I, losses occur during dispersal in autumn and spring when food supply is not critical and the form of agonistic behavior is territory defense. In Phase II, losses occur from food shortage in late winter when food supply is of critical importance and the form of agonistic behavior involves dominance rank. Brown has pointed out that the number in the whole population of the species is unaffected by dispersal alone and that it is the second phase that is more sensitive to the degree of overpopulation and is more efficient as a regulator. In the following sections, I demonstrate that Phases I and II should be treated together. I propose that the dispersal phase is also regulated by dominance rank and that the winter distribution of members of a species is a result of the dispersal phase.

Dispersal behavior in vertebrates can be of two types (Howard, 1960): (1) innate dispersal, in which animals are predisposed at birth to disperse beyond the confines of their parental home range, ignoring available and suitable niches and dispersing into strange and sometimes unfavorable habitats, and (2) environmental dispersal, the movement an animal makes away from its birthplace in response to crowded conditions (mate selection, territoriality, lack of suitable homesites and resources, or parental rejection). Environmental dispersers have an inherited homing ability. Innate dispersal is density-independent, according to Howard, while environmental dispersal is, of course, density-dependent, and both are thought to be inheritable. Social dominance can help explain the dichotomy between innate and environmental dispersal. Dominant individuals satisfy Howard's criteria for an environmental disperser in that they tend to be more sedentary and move only short distances, provided resources are available. Subordinate individuals act more like innate dispersers in that they disperse even in the absence of high population pressures, because they are displaced or excluded by dominants well before food resources become limited.

Two qualitatively distinguishable kinds of dispersal that result in two corresponding kinds of emigration have been proposed by Lidicker (1962, 1975). One type he calls "saturation dispersal" and the other "pre-saturation dispersal." Lidicker has suggested that saturation emigration is the outward movement of surplus individuals from a population living at or near its carrying capacity. Such individuals showing saturation dispersal behavior are faced with the immediate choice of staying in the population and almost certainly dying or emigrating and probably dying. Lidicker believes that the animals in this case are social outcasts, juveniles, very old individuals, those in poor condition, and in general those least able to cope with local conditions. In pre-saturation emigration, animals move from the population before their habitat becomes saturated with individuals. Pre-saturation emigrants may be characterized by possessing a particular sensitivity to increasing densities, or they may have discovered some better home location during an exploratory excursion. Once again, I believe that behavioral dominance can explain the different types of dispersal proposed by Lidicker. The differential dispersal with regard to age and sex strongly suggests that a form of behavioral dominance is in operation. DeLong (1967) has suggested that socially dominant feral house mice (*Mus musculus*) are less likely to leave their home area and are able to resist the invasion of outsiders.

Dominance hierarchies can be important in maintaining homeostasis by eliminating subordinates when food is reduced and by excluding individuals of lower rank from breeding (Wynne-Edwards, 1962, p. 143). There is general agreement that social pressures drive dispersal long before economic saturation of the habitat is reached (see Lidicker, 1975, p. 108), and Christian (1970) has shown that social hierarchies constitute a major force in the differential emigration of adults and young mammals. In starlings (*Sturnus vulgaris*), there is a social rank in the wild, but the subordinates are present only temporarily and soon emigrate (Davis, 1959). Murton *et al.* (1966) and Murton (1968) have shown that subordinate wood pigeons (*Columba palumbus*) accumulate less food during the day than do dominants. When food density is high, both dominants and subordinates can obtain sufficient food to meet their energy demands, but the rate of food collection is lower for the subordinate (Fig. 3). When food density becomes limited (seasonal depletion), only the dominant may obtain sufficient food to survive. What, then, is the fate of the subordinate? Most animals that have restricted access to food increase locomotor activity (Wald and Jackson, 1944; Bare, 1959; Reynierse *et al.*, 1972; Hughes and Wood-Gush, 1973). Furthermore, several studies on rodents show that brief depri-

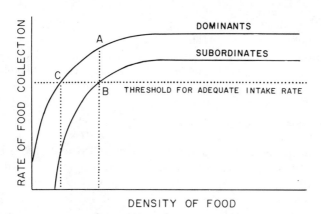

DENSITY OF FOOD

Fig. 3. Feeding rates of dominant and subordinate wood pigeons in relation to food density. The curves show the numbers of cereal grains eaten (pecks per minute) by birds in various densities of food as counted in field observations. The density of food below which an adequate intake rate cannot be maintained is given by the vertical line below B for a subordinate and below C for a dominant. (After Murton, 1968.)

vation produces larger activity increases in younger animals than in older animals (Campbell *et al.*, 1961; Finger, 1962; Hodge *et al.*, 1967). The increased locomotor activity is highly adaptive and enables the subordinate to seek relief from the stressful situation by emigration or dispersal. This process can result in (1) finding an area free of dominants and of the same quality with respect to food; (2) finding an area free of dominants but of reduced quality with respect to food; or (3) not finding an area where food can be obtained and perishing. In most cases option 1 is not likely for a mobile species at or near carrying capacity. Options 2 and 3 are more likely to be found. In option 3, the subordinates are eliminated from the population. In option 2, the subordinates settle in areas that are of reduced quality and hence free of dominants (provided the dominants have enough quality areas) and are themselves "dominant" in the new areas. The resultant distribution of dominants and subordinates in relation to the quality of habitats tends to minimize intraspecific competition and to maximize exploitation of food resources when resources are reduced. As food becomes further reduced, there may be an additional shift so that the most subordinate individuals have to move to habitats that are quite marginal and where their survival is in serious jeopardy (Fretwell, 1969). Movement in response to decreased food resources may have additional costs above the energetic investment. Glickman (1971) has demonstrated in a laboratory study that when groups of mice were caged with a barred owl (*Strix varia*), it was the more active mice that were captured and eaten. Lidicker (1975) has also discussed the higher risk of death from lack of shelter, starvation, and predation in dispersing small mammals. Subordinates on the periphery of a group are more susceptible to predation than dominants in the center (Coulson, 1968; Hamilton, 1971).

If dominance rank influences dispersal, then it must certainly play an important role in irruptions and partial, short-, and long-distance migrations, for they show fundamental similarities to dispersal (emigration and immigration). Howard (1960) and Lidicker (1962, 1975) have discussed various aspects of dispersal, migration, and nomadism. Short-distance migration and partial migration (Lack, 1944) are strikingly similar to environmental dispersal and irruptive movements. Ulfstrand (1963) has commented on similarities between irruptions and migrations, and Newton (1970) has further elaborated on the traits that unify dispersal, emigration, and true migration. Heape (1931), Lack (1954), and Berndt and Sternberg (1968) have discussed all the forms of dispersal and migratory behavior and have shown that they are closely related and often confused. All occur in response to resource limitation (exploitative and interference competition; see Miller, 1967), crowding, or both, and all can have dramatic effects on gene flow.

Dispersal and migration behavior collectively are often strongly biased with respect to age and sex. In general, young show a tendency to travel farther than adults (Lack, 1943; Kluyver and Tinbergen, 1953; Lidicker, 1962, 1975; Perrins, 1963; Dixon and Gilbert, 1964; Pinowski, 1965; Sadleir, 1965; Healey, 1967; Reimer and Petras, 1967; Berndt and Sternberg, 1969; Rheinwald and Gutscher, 1969; Gaydar, 1973; Baker, 1975; Branch, 1975; Balát, 1976; Picozzi and Weir, 1976; but see King, 1955). Likewise, females tend to disperse more than males (Lack, 1943, 1954; Kluyver, 1957; Lidicker 1962, 1975; Delius, 1965; Rowley, 1965; Berndt and Sternberg, 1969; Catchpole, 1972; Bulmer, 1973; Gaydar, 1973; Zahavi, 1974; Greenwood and Harvey, 1976).

In a summary of current knowledge regarding the determinants of dominance, Wilson (1975) concluded that "adults are dominant over juveniles, and males are usually dominant over females. In multimale societies, it is typical for the rank ordering of the males to lie entirely above that of the females, or at most overlap it slightly. In such cases juvenile males sometimes work their way up through the female hierarchy before achieving greater than omega status with reference to the males" (p. 291). In some cases, females are dominant over males (Simmons, 1970; Rowell, 1971; Kruuk, 1972). Thompson (1960) and Snyder and Wiley (1976) have reviewed several cases where female birds are dominant over males, and Ralls (1976, p. 264) has examined this matter in mammals. Large size and familiarity with the area can also confer dominance (Allee et al., 1949, p. 413). Fretwell (1969) has shown that larger birds are dominant over smaller birds. A number of field studies have shown that familiarity with a given site confers dominance (Colquhoun, 1942; Brian, 1949; Brown, 1963; Dixon, 1965; Glase, 1973; Smith, 1976).

If dominance is a major factor in postbreeding dispersal, irruptions, and partial and short-distance migrations, and if young tend to move farther away from the breeding grounds than adults and females farther than males, provided males are dominant, then during the nonbreeding season there should be a partial, orderly segregation of age and sex classes throughout the nonbreeding range. In other words, dominance order in the wintering range can be measured on the basis of the distance of each individual from its breeding area. There is abundant evidence that this is the case, and it has been verified by the use of feeding trays in largely sedentary species (Colquhoun, 1942; Odum, 1942; Brian, 1949). Gauthreaux (1975) has given numerous examples in various vertebrates, and Ketterson and Nolan (1976) have summarized the differential sex and age ratios of birds throughout their winter ranges. Interestingly, a similar phenomenon is also found in montane avifaunas. The change in population structure with increasing altitude is typically that immatures are found at the bottom of the altitudinal range; somewhat higher, immatures plus birds in adult plumage but in nonbreeding condition are encountered, with females usually appearing at lower altitudes than males; next comes the optimal part of the species' range, with singing males and adults of both sexes in breeding condition; and finally, another band of immatures but of few adults appears at the upper altitudinal limit in some species (J. M. Diamond, 1972). This distribution tendency of immatures to be found at the peripheries of the altitudinal range characterizes most species of the New Guinea montane avifauna and not just those that belong to altitudinal sequences (J. M. Diamond, 1972). A. W. Diamond (1973) has also commented on a similar altitudinal variation in a resident and a migrant passerine in Jamaica. Dixon and Gilbert (1964) have reported that in chickadees (*Parus*), the young show more altitudinal migration than adults. In cases where both sexes and all age classes winter in the same general area, there is evidence to document that the males, if they are dominant, have the better habitat and that adults have a better habitat than young (Nelson and Janson, 1949; Kluyver, 1957; Fretwell, 1969; Nisbet and Medway, 1972).

With regard to the timing of migration, one would suppose that the dominant would be the first to arrive on the breeding grounds to establish and defend a territory and that the subordinates would arrive later. There is overwhelming support for this idea in the literature. In all cases where males are dominant, they arrive on the breeding grounds ahead of the females, and older adults arrive before younger individuals (see Gauthreaux, 1975, for a review of this subject). In cases where females are dominant, they arrive on the breeding grounds first (Thompson, 1960; Höhn, 1967; Kistchinski, 1975) and may play a major role in territorial selection and defense. There is also evidence that the earliest arrivals on the breeding grounds get the best breeding areas (Carrick and Ingham, 1967; Coulson, 1968; Haneda and Teranishi, 1968; Spurr, 1972, 1974). Murray (1967) considered a hypothetical population in order to gain insight into the factors that were responsible for the skewed distributions of dispersal distances reported in vertebrates. In his population, the most effective factor was the relative dominance of the individuals in procuring breeding sites. Murray postulated that if this were true in natural populations, dispersal would permit each individual to maximize its chance to reproduce. He further concluded that the earlier explanations emphasizing selection for genotypes that were advantageous to the species but disadvantageous to individuals are not necessary to explain the skewed distributions of dispersal distances in vertebrate populations. Oakeshott (1974) has shown for *Mus musculus* that relative weights and initial orders of arrival influence social dominance. It would appear that priority of arrival provides priority of access. Thus, the dominant early arrivals have a greater chance for reproductive success and can breed earlier with the most dominant females as soon as they arrive on the breeding grounds. This would suggest that the older, more experienced individuals would have greater reproductive success than the younger individuals, which is generally held to be true (Lack, 1966; Brown, 1969; Watson and Moss, 1970). Early breeding by dominant adults would also confer an advantage on their young in terms of subsequent pressures to disperse. Young produced later in the mating season are usually subordinate and consequently are the first to disperse (Michener and Michener, 1935; Erickson, 1938; Kluyver and Tinbergen, 1953; Dixon 1956; Perrins, 1963; Pinowski, 1965). Thus, the conditions on the breeding grounds can dictate and influence postbreeding events (dispersal, emigration, and migration). These in turn can influence population dynamics during the nonbreeding season (differential mortality, sex ratios, age classes), which can regulate the return to the breeding grounds (dispersal, immigration, migration, and homing) and act as a feedback on the population during the breeding season (mating, nesting, and reproductive success). For instance, dominance rank can influence the mating system by affecting the sex ratio of birds during the nonbreeding season and by dictating whether adult males arrive on the breeding ground before or after the adult females (Gowaty and Gauthreaux, 1977).

Although homing behavior has received considerable attention in the literature, few workers have emphasized the ecological adaptiveness of this behavior. It is particularly appropriate to do so now. The advantages of returning to an area where breeding was previously successful are obvious. It follows that the degree and accuracy of prebreeding homing (*Ortstreue*) should vary depending on the age of the individuals and perhaps on their sex. The data showing greater dispersal of young than adults on the breeding grounds support the notion that adults are better at homing, but what about sexual differences? This data base is meager, but it suggests that in territorial species that return to their breeding site, adult females show less breeding-site attachment than adult males and nest farther away from areas where they have previously bred (Delius, 1965; Rowley, 1965; Berndt and Sternberg, 1969; Catchpole, 1972; Zahavi, 1974). Thus, an individual's rank in the dominance hierarchy may also be directly related to its homing ability during or shortly following its movement toward the breeding area. Such variability in the homing response is certainly adaptive in that it permits some degree of gene flow and prevents inbreeding. Similarly, experienced adults have been shown to return to previously established nonbreeding sites (Farner, 1945; Lack, 1954; Perdeck, 1958; Löhrl, 1959; Schwartz, 1963; Backhurst, 1972; DeSmet, 1972; Flegg and Cox, 1972; Diamond and Smith, 1973; Imboden, 1974; Ralph and Mewaldt, 1975; Mewaldt, 1976).

Lack (1968) has emphasized that factors acting in the winter quarters are critical in regulating populations of migratory species. Fretwell (1972) has echoed this emphasis. Lack (1968) has stated that the critical factor is not the absolute quantity of food but the amount available in relation to actual or potential competition. There have been and there continue to be strong selective pressures for dispersal behaviors (ranging from true dispersal to long-distance migration) so that the young minimize competition with adults and females minimize competition with males and yet manage to settle in the nonbreeding season in habitats or geographical areas where they can survive. Within and between species, behavioral dominance (or the dominance gradient) is a major means of achieving this objective. Dominance rank ultimately reduces competition through a partial segregation of ages and sexes either geographically or by habitat and serves as a regulatory mechanism by damping fluctuations in population density. Similarly, at the interspecific level, behavioral dominance can enhance the process of adaptive radiation.

VI. SUMMARY

Dominance hierarchies are distributed throughout the animal kingdom, albeit in a highly irregular fashion. Considerable attention has been directed to the numerous and confusing behavioral definitions of social dominance and to the means of measuring it. The types of behavior used to judge dominance have been nearly as diverse as the species observed, and the concept has been applied to a wide range of groups varying in size from a few individuals to two or more competitive species. In this paper, I have stressed the ecological and evolutionary utility of the concept of dominance at the population level, and I have constructed a model that interrelates dominance rank, the changing abundance of a limiting resource, and dispersal behavior in terms of habitat quality and geographical distance. It is suggested that dominance rank influences dispersal, irruptions, short- and long-distance migrations, and homing. Invoking dominance rank as an underlying regulatory factor in the population dynamics of a species helps explain differential habitat utilization or geographic distribution during the nonbreeding season, differential mortality during the nonbreeding and breeding seasons, the orderly return of individuals to the breeding grounds or the breeding habitat, the establishment and the quality of the individual's territory, mating success, and ultimately the individual's reproductive success. Behavioral dominance is viewed as a means of reducing intraspecific and interspecific competition and costly interactions by permitting a group of individuals (or species) to coexist in an orderly fashion with minimal dele-terious interactions while permitting the most efficient utilization of resources. Unlike adaptive radiation, the individuals (or species) are not locked into an exploitation strategy and can change their strategy depend-ing on competitive circumstance. In this regard, social dominance is a plastic strategy much the same as learning with profound ecological implications for the individual, the class, the group, and ultimately the species.

VII. ACKNOWLEDGMENTS

This paper is the product of continuous and helpful stimulation from my graduate students in behavioral ecology. Their questions and discussions contributed substantially to my ability to conceptualize behavioral domi-nance in ecological and evolutionary terms. I also benefited greatly from discussions with and comments from H. Wiley, K. P. Able, H. H. Shugart, S. T. Emlen, and D. H. Morse. F. R. Moore kindly drew the figures, and P. Gowaty and B. Alexander were particularly helpful throughout the prepara-tion of the manuscript. A. Snider and P. Hamel were indispensable in typ-ing and proofreading the final manuscript.

VIII. REFERENCES

Aikman, D., and Hewitt, G. (1972). An experimental investigation of the rate and form of dis-persal in grasshoppers. *J. Appl. Ecol.* **9**:807–817.
Alexander, B. K., and Hughes, J. (1971). Canine teeth and rank in Japanese monkeys (*Macaca fuscata*). *Primates* **12**:91–93.
Alexander, R. D. (1961). Aggressiveness, territoriality, and sexual behavior in field crickets (Orthoptera: Gryllidae). *Behaviour* **17**:130–223.
Alexander, R. D., and Thomas, E. S. (1959). Systematic and behavioral studies on the crickets of the *Nemobius fasciatus* group (Orthoptera: Gryllidae, Nemobiinae). *Ann. Entomol. Soc. Am.* **52**:591–605.
Alexander, T. R. (1964). Observations on the feeding behavior of *Bufo marinus* (Linné). *Herpetologica* **20**:255–259.
Allee, W. C. (1939). *The Social Life of Animals,* Norton, New York, 293 pp.
Allee, W. C. (1942). Social dominance and subordination among vertebrates. *Biol. Symp.* **8**:139–162.
Allee, W. C., and Collias, N. E. (1938). Effects of injections of ephinephrine on the social order in small flocks of hens. *Anat. Rec. Suppl.* **72**:119.

Allee, W. C., and Collias, N. E. (1940). The influence of estradiol on the social organization of flocks of hens. *Endocrinology* **27**:87–94.

Allee, W. C., and Dickinson, J. C., Jr. (1954). Dominance and subordination in the smooth dogfish *Mustelus canis* (Mitchill). *Physiol. Zool.* **27**:356–364.

Allee, W. C., and Douglis, M. B. (1945). A dominance order in the hermit crab, *Pagurus longicarpus* Say. *Ecology* **26**:411–412.

Allee, W. C., Collias, N. E., and Lutherman, C. Z. (1939). Modification of the social order in flocks of hens by the injection of testosterone propionate. *Physiol. Zool.* **12**:412–440.

Allee, W. C., Collias, N. E., and Beeman, E. (1940). The effect of thyroxin on the social order in flocks of hens. *Endocrinology* **27**:827–835.

Allee, W. C., Emerson, A. E., Park, O., Park, T., and Schmidt, K. (1949). *Principles of Ecology,* W. B. Saunders, Philadelphia, 837 pp.

Altmann, M. (1952). Social behaviour of elk, *Cervus canadensis n.,* in the Jackson Hole of Wyoming. *Behaviour* **4**:116–143.

Altmann, S. A. (1962). A field study of the sociobiology of rhesus monkeys, *Macaca mulatta. Ann. N.Y. Acad. Sci.* **102**:330–435.

Antonov, V. V. (1976). On the mechanisms controlling the development of hierarchical relationships in dogs. *Zh. Obshch. Biol.* **37**:310–313.

Archer, J. (1970). Effects of population density on behaviour in rodents. In Crook, J. H. (ed.), *Social Behaviour in Birds and Mammals: Essays on the Social Ethology of Animals and Man,* Academic Press, New York, pp. 169–210.

Backhurst, G. C. (1972). East African bird ringing report 1970–71. *J. East Afr. Nat. Hist. Soc. Natl. Mus.* No. 136, 16 pp.

Baenninger, L. (1966). The reliability of dominance orders in rats. *Anim. Behav.* **14**:367–371.

Baenninger, L. (1970). Social dominance orders in the rat: "Spontaneous" food and water competition. *J. Comp. Physiol. Psychol.* **71**:202–209.

Baird, R. C. (1968). Aggressive behaviour and social organization in *Mollienesia latipinna* LeSueur. *Texas J. Sci.* **20**:157–176.

Baker, A. J. (1975). Age structure and sex ratio of live-trapped samples of South Island pied oystercatchers (*Haematopus ostralegus finschi*). *Notornis* **22**:189–194.

Balát, F. (1976). Dispersionsprozesse und Brutortstreue beim Feldsperling *Passer montanus. Zool. Listy* **25**:39–49.

Baldwin, J. D. (1971). The social organization of a semifree-ranging troop of squirrel monkeys (*Saimiri sciureus*). *Folia Primatol.* **14**:23–50.

Banks, E. M. (1956). Social organization in red jungle fowl hens (*Gallus gallus* subsp.). *Ecology* **37**:239–248.

Bare, J. K. (1959). Hunger, deprivation, and the day–night cycle. *J. Comp. Physiol. Psychol.* **52**:129–131.

Barlow, G. W. (1974). Contrasts in social behavior between Central American cichlid fishes and coral-reef surgeon fishes. *Am. Zool.* **14**:9–34.

Barnett, S. A. (1963). *The Rat: A Study in Behaviour,* Aldine, Chicago, 288 pp.

Barnett, S. A. (1967). *A Study in Behaviour,* Methuen, London, 304 pp.

Bartlett, D. P., and Meier, G. W. (1971). Dominance status and certain operants in a communal colony of rhesus monkeys. *Primates* **12**:209–219.

Becker, G., and Flaherty, T. B. (1968). Group size as a determinant of dominance-hierarchy stability in the rat. *J. Comp. Physiol. Psychol.* **66**:473–476.

Beebe, W. (1947). Notes on the hercules beetle, *Dynastes hercules* (Linn.), at Rancho Grande, Venezuela, with special reference to combat behavior. *Zoologica* **32**:109–116.

Bennett, M. A. (1939). The social hierarchy in ring doves. *Ecology* **20**:337–357.

Bennett, M. A. (1940). The social hierarchy in ring doves. II: The effect of treatment with testosterone proprionate. *Ecology* **21**:148–165.

Berndt, R., and Sternberg, H. (1968). Terms, studies and experiments on the problems of bird dispersion. *Ibis* **110**:256–269.

Berndt, R., and Sternberg, H. (1969). Alters- und Geschlechtsunterschiede in der Dispersion des Trauerschnäppers (*Ficedula hypoleuca*). *J. Ornithol.* **110**:256–269.

Bernstein, I. S. (1970). Primate status hierarchies. In Rosenblum, L. A. (ed.), *Primate Behaviour: Developments in Field and Laboratory Research*, Vol. 1, Academic Press, New York, pp. 71–109.

Bernstein, I. S. (1976). Dominance, aggression and reproduction in primate societies. *J. Theoret. Biol.* **60**:459–472.

Blaustein, A. R., and Risser, A. C., Jr. (1976). Interspecific interactions between three sympatric species of kangaroo rats (*Dipodomys*). *Anim. Behav.* **24**:381–385.

Boice, R., and Witter, D. W. (1969). Hierarchical feeding behaviour in the leopard frog (*Rana pipiens*). *Anim. Behav.* **17**:474–479.

Boice, R., Quanty, C. B., and Williams, R. C. (1974). Competition and possible dominance in turtles, in toads, and in frogs. *J. Comp. Physiol. Psychol.* **86**:1116–1131.

Boreman, J., and Price, E. (1972). Social dominance in wild and domestic Norway rats (*Rattus norvegicus*). *Anim. Behav.* **20**:534–542.

Bovbjerg, R. V. (1953). Dominance order in the crayfish *Orconectes virilis* (Hagen). *Physiol. Zool.* **26**:173–178.

Bovbjerg, R. V. (1956). Some factors affecting aggressive behavior in crayfish. *Physiol. Zool.* **29**:127–136.

Boyd, H. (1953). On encounters between wild white-fronted geese in winter flocks. *Behaviour* **5**:85–129.

Braddock, J. C. (1945). Some aspects of the dominance–subordination relationship in the fish *Platypoecilus maculatus*. *Physiol. Zool.* **18**:176–195.

Braddock, J. C. (1949). The effect of prior residence upon dominance in the fish *Platypoecilus maculatus*. *Physiol. Zool.* **22**:161–169.

Branch, G. M. (1975). Mechanisms reducing intraspecific competition in *Patella* spp.: Migration, differentiation and territorial behaviour. *J. Anim. Ecol.* **44**:575–600.

Braun, R. (1958). Das Sexualverhalten der Krabbenspinne *Diaea dorsata* (F.) und der Zartspinne *Anyphaena accentuata* (Walck.) als Hinweis auf ihre systematische Eingliederung. *Zool. Anz.* **160**:119–134.

Brian, A. D. (1949). Dominance in the great tit *Parus major*. *Scott. Nat.* **61**:144–155.

Brian, M. V. (1952). The structure of a dense natural ant population. *J. Anim. Ecol.* **21**:12–24.

Brown, J. L. (1963). Aggressiveness, dominance and social organization in the Steller jay. *Condor* **65**:460–484.

Brown, J. L. (1969). Territorial behavior and population regulation in birds: A review and re-evaluation. *Wilson Bull.* **81**:293–329.

Brown, J. L. (1975). *The Evolution of Behavior*, Norton, New York, 761 pp.

Bulmer, M. G. (1973). Inbreeding in the great tit. *Heredity* **30**:313–325.

Calhoun, J. B. (1962). The ecology and sociology of the Norway rat. U.S. Department of Health, Education, and Welfare, Public Health Service Document No. 1008, 288 pp.

Campbell, B. A., Teghtsoonian, R., and Williams, R. A. (1961). Activity, weight loss, and survival time of food deprived rats as a function of age. *J. Comp. Physiol. Psychol.* **54**:216–219.

Carpenter, C. R. (1942a). Sexual behavior of free-ranging rhesus monkeys (*Macaca mulatta*). I: Specimens, procedures and behavioral characteristics of estrus. *J. Comp. Psychol.* **33**:113–142.

Carpenter, C. R. (1942b). Sexual behavior of free-ranging rhesus monkeys (*Macaca mulatta*). II: Periodicity of estrus, homosexual, autoerotic and nonconformist behavior. *J. Comp. Psychol.* 33:143-162.

Carrick, R., and Ingham, S. E. (1967). Antarctic sea-birds as subjects for ecological research. *Jap. Antarct. Res. Exped. Sci. Repts.*, Special Issue, 1:151-184.

Catchpole, C. K. (1972). A comparative study of territory in the reed warbler (*Acrocephalus scirpaceus*) and sedge warbler (*A. schoenobaenus*). *J. Zool.* 166:213-231.

Chalmers, N. R., and Rowell, T. E. (1971). Behaviour and female reproductive cycles in a captive group of mangabeys. *Folia Primatol.* 14:1-14.

Chance, M. R. A., and Jolly, C. J. (1970). *Social Groups of Monkeys, Apes and Men,* E. P. Dutton, New York, 224 pp.

Chase, I. D. (1974). Models of hierarchy formation in animal societies. *Behav. Sci.* 19:374-382.

Christian, J. J. (1970). Social subordination, population density, and mammalian evolution. *Science* 168:84-90.

Christian, J. J. (1971). Fighting, maturity and population density in *Microtus. J. Mammal.* 52:556-567.

Collias, N. E. (1943). Statistical analysis of factors which make for success in initial encounters between hens. *Am. Nat.* 77:519-538.

Collias, N. E. (1944). Aggressive behavior among vertebrates. *Physiol. Zool.* 17:83-123.

Collias, N. E., and Collias, E. C. (1967). A field study of the red jungle fowl in north-central India. *Condor* 69:360-386.

Collias, N. E., and Jahn, L. R. (1959). Social behavior and breeding success in Canada geese (*Branta canadensis*) confined under seminatural conditions. *Auk* 76:478-509.

Collias, N. E., and Taber, R. E. (1951). A field study of some grouping and dominance relations in ring-necked pheasants. *Condor* 53:265-275.

Colquhoun, M. K. (1942). Notes on the social behaviour of blue tits. *Br. Birds* 35:234-240.

Coulson, J. C. (1968). Differences in the quality of birds nesting in the centre and on the edges of a colony. *Nature* 217:478-479.

Crane, J. (1957). Imaginal behavior in butterflies of the family Heliconnidae: Changing social patterns and irrelevant actions. *Zoologica* 42:135-145.

Darling, F. F. (1937). *A Herd of Red Deer,* Oxford University Press, London, 215 pp.

Davis, D. E. (1942). The phylogeny of social nesting habits in the Crotophaginae. *Q. Rev. Biol.* 17:115-134.

Davis, D. E. (1958). The role of density in aggressive behaviour in the house mouse. *Anim. Behav.* 6:207-211.

Davis, D. E. (1959). Territorial rank in starlings. *Anim. Behav.* 7:214-221.

DeFries, J. C., and McClearn, G. E. (1970). Social dominance and Darwinian fitness in the laboratory mouse. *Am. Nat.* 104:408-411.

Deleurance, É.-P. (1948). Le comportement reproducteur est indépendant de la présence des ovaires chez *Polistes* (Hyménoptères Vespides). *C. R. Acad. Sci., Paris* 227:866-867.

Deleurance, É.-P. (1952). Le polymorphisme social et son déterminisme chex le Guêpes. In Grassé, P.-P. (ed.), *Structure et Physiologie des Sociétés Animales,* Int. Ctr. Nat. Rech. Sci., Paris, pp. 141-155.

Delius, J. D. (1965). A population study of skylarks, *Alauda arvensis. Ibis* 107:446-492.

DeLong K. T. (1967). Population ecology of feral house mice. *Ecology* 48:611-634.

Den Boer, P. J. (1968). Spreading of risk and stabilization of animal numbers. *Acta Biotheor.* 18:165-194.

DeSmet, W. M. A. (1972). The migration of the cuckoo (*Cuculus canorus*). *Gerfaut* 62:277-305.

DeVore, B. I. (1971). The evolution of human society. In Eisenberg, J. F., and Dillon, W. S. (eds.), *Man and Beast: Comparative Social Behavior,* Smithsonian Institution Press, Washington, pp. 297–311.

Diamond, A. W. (1973). Altitudinal variation in a resident and a migrant passerine on Jamaica. *Auk* **90**:610–618.

Diamond, A. W., and Smith, R. W. (1973). Returns and survival of banded warblers wintering in Jamaica. *Bird-Banding* **44**:221–224.

Diamond, J. M. (1972). Avifauna of the eastern highlands of New Guinea. *Publ. Nuttall Ornith. Club,* No. 12, Cambridge, Mass., 438 pp.

Dice, L. R., and Howard, W. E. (1951). Distance of dispersal by prairie deermice from birthplaces to breeding sites. *Univ. Mich. Contrib. Lab. Vert. Biol.* **50**:1–15.

Dingle, H. (1965). The relation between age and flight activity in the milkweed bug, *Oncopeltus. J. Exp. Biol.* **42**:269–283.

Dingle, H. (1966). Some factors affecting flight activity in the individual milkweed bug (*Oncopeltus*). *J. Exp. Biol.* **44**:335–343.

Dingle, H. (1968). The influence of environment and heredity on flight activity in the milkweed bug *Oncopeltus. J. Exp. Biol.* **48**:175–184.

Dixon, K. L. (1956). Territoriality and survival in the plain titmouse. *Condor* **58**:169–182.

Dixon, K. L. (1963). Some aspects of the social organization of the Carolina chickadee. *Proc. 13th Int. Ornithol. Congr.* **1**:240–258.

Dixon, K. L. (1965). Dominance–subordination relationships in the mountain chickadee (*Parus gambeli*). *Condor* **67**:291–299.

Dixon, K. L., and Gilbert, J. D. (1964). Altitudinal migration in the mountain chickadee. *Condor* **66**:61–64.

Dobzhansky, T., and Wright, S. (1943). Genetics of natural populations. X: Dispersion rates in *Drosophila pseudoobscura. Genetics* **28**:304–340.

Douglis, M. B. (1946). Some evidence of a dominance subordinance relationship among lobsters, *Homarus americanus. Anat. Rec.* **94**:57.

Drews, D. R. (1973). Group formation in captive *Galago crassicaudatus*: Notes on the dominance concept. *Z. Tierpsychol.* **32**:425–435.

Drews, D. R., and Dickey, C. L. (1977). Observational and competitive measures of dominance in rats. *Psychol. Record* **27**:331–338.

Drickamer, L. C., and Vandenbergh, J. G. (1973). Predictors of social dominance in the adult female golden hamster (*Mesocricetus auratus*). *Anim. Behav.* **21**:564–570.

Drickamer, L. C., Vandenbergh, J. G., and Colby, D. R. (1973). Predictors of dominance in the male golden hamster (*Mesocricetus auratus*). *Anim. Behav.* **21**:557–563.

Dunbar, R. I. M., and Crook, J. H. (1975). Aggression and dominance in the weaverbird, *Quelea quelea. Anim. Behav.* **23**:450–459.

Dunbar, R. I. M., and Dunbar, E. P. (1976). Contrasts in social structure among black-and-white colobus monkey groups. *Anim. Behav.* **24**:84–92.

Dunham, D. W. (1966). Agonistic behavior in captive rose-breasted grosbeaks, *Pheucticus ludovicianus* (L.). *Behaviour* **27**:160–173.

Eberhard, M. J. W. (1969). The social biology of polistine wasps. *Misc. Publs., Mus. Zool., Univ. Michigan,* Ann Arbor **140**:1–101.

Eisenberg, J. F. (1963). The intraspecific social behavior of some cricetine rodents of the genus *Peromyscus. Am. Midland Nat.* **69**:240–246.

Eisenberg, J. F., and Kuehn, R. E. (1966). The behavior of *Ateles geoffroyi* and related species. *Smithsonian Misc. Coll.* **151**, 63 pp.

Erickson, J. G. (1967). Social hierarchy, territoriality, and stress reactions in sunfish. *Physiol. Zool.* **40**:40–48.

Erickson, M. M. (1938). Territoriality, annual cycle, and numbers in a population of wren-tits (*Chamaea fasciata*). *Univ. Calif. Publ. Zool.* **42**:247–334.

Estes, R. D. (1974). Social organization of the African Bovidae. In Geist, V., and Walther, F. (eds.), *The Behaviour of Ungulates and Its Relation to Management*, Int. Union Cons., Morges, pp. 166–205.

Evans, L. T. (1936). A study of social hierarchy in the lizard, *Anolis carolinensis. J. Genet. Psychol.* **48**:88–111.

Ewbank, R., Meese, G. B., and Cox, J. E. (1974). Individual recognition and the dominance hierarchy in the domesticated pig. The role of sight. *Anim. Behav.* **22**:473–480.

Ewing, L. S., and Ewing, A. W. (1973). Correlates of subordinate behaviour in the cockroach, *Nauphoeta cinerea. Anim. Behav.* **21**:571–578.

Farentinos, R. C. (1972). Social dominance and mating activity in the tassel-eared squirrel (*Sciurus aberti ferreus*). *Anim. Behav.* **20**:316–326.

Farner, D. S. (1945). The return of robins to their birthplaces. *Bird-Banding* **16**:81–99.

Farr, J. A. III, and Herrnkind, W. F. (1974). A quantitative analysis of social interaction of the guppy, *Poecilia reticulata* (Pisces: Poeciliidae) as a function of population density. *Anim. Behav.* **22**:582–591.

Fielder, D. R. (1965). A dominance order for shelter in the spiny lobster *Jasus lalandei* (H. Milne-Edwards). *Behaviour* **23**:236–245.

Figler, N. H., Klein, R. M., and Peeke, H. V. S. (1976). Establishment and reversibility of dominance relationships in jewel fish, *Hemichromis bimaculatus* Gill (Pisces, Cichlidae). Effects of prior exposure and prior residence situations. *Behaviour* **58**:254–271.

Finger, F. W. (1962). Activity change under deprivation as a function of age. *J. Comp. Physiol. Psychol.* **55**:100–102.

Fischel, W. (1927). Beiträge zur Sociologie des Haushuhns. *Biol. Zentrabl.* **47**:678–695.

Flegg, J. J. M., and Cox, C. J. (1972). Movement of black-headed gulls from colonies in England and Wales. *Bird Study* **19**:228–240.

Fraser, A. F. (1968). *Reproductive Behaviour in Ungulates*, Academic Press, New York, 202 pp.

Free, J. B. (1955). The division of labour within bumblebee colonies. *Insectes Sociaux* **2**:195–212.

Free, J. B. (1961). The social organization of the bumble-bee colony. A lecture given to the Central Association of Bee-keepers on 18th January 1961. North Hants Printing and Publishing Co., Fleet, Hants., England, 11 pp.

Free, J. B., and Butler, C. G. (1959). *Bumblebees*, Collins, London, 208 pp.

Fretwell, S. D. (1969). Dominance behavior and winter habitat distribution in juncos (*Junco hyemalis*). *Bird-Banding* **40**:1–25.

Fretwell, S. D. (1972). Populations in a seasonal environment. *Monogr. Popul. Biol.*, No. 5, Princeton University Press, Princeton, N.J., 217 pp.

Froese, A. D., and Burghardt, G. M. (1974). Competition in captive juvenile snapping turtles, *Chelydra serpentina. Anim. Behav.* **22**:735–740.

Gartlan, J. S. (1968). Structure and function in primate society. *Folia Primatol.* **8**:89–120.

Gauthreaux, S. A., Jr. (1975). Diffusion theory, dispersal, and migration. Unpublished manuscript, 55 pp.

Gaydar, A. A. (1973). Ringing of *Tetrastes bonasia* and its results. *Byull. Mosk. O.-Va. Ispyt. Prir. Otd. Biol.* **78**:120–124.

Geist, V. (1971). *Mountain Sheep: A Study in Behaviour and Evolution*, University of Chicago Press, Chicago, 383 pp.

Gervet, J. (1956). L'action des températures différentielles sur la monogynie fonctionnelle chez les *Polistes* (Hyménoptères Vespides). *Insectes Sociaux* **3**:159–176.

Gervet, J. (1962). Étude de l'effet de groupe sur la ponte dans la société polygyne de *Polistes gallicus*. *Insectes Sociaux* **9**:231–263.

Gibb, J. A. (1960). Populations of tits and goldcrests and their food supply in pine populations. *Ibis* **102**:163–208.

Gibb, J. A. (1962). The importance of territory and food supply in the natural control of a population of birds. *Sci. Rev. Australia* **20**:20–21.

Gibson, R. N. (1968). The agonistic behaviour of juvenile *Blennus pholis* L. (Teleosti). *Behaviour* **30**:192–217.

Gilbert, L. E., and Singer, M. C. (1973). Dispersal and gene flow in a butterfly species. *Am. Nat.* **107**:58–72.

Ginsburg, B., and Allee, W. C. (1942). Some effects of conditioning on social dominance and subordination in inbred strains of mice. *Physiol. Zool.* **15**:485–506.

Glase, J. C. (1973). Ecology of social organization in the black-capped chickadee. *Living Bird* **12**:235–267.

Glickman, S. E. (1971). Curiosity has killed more mice than cats. *Psychology Today* **5**:54–56, 86.

Goforth, W. R., and Baskett, T. S. (1971). Social organization of penned mourning doves. *Auk* **88**:528–542.

Goodwin, D. (1951). Some aspects of the behaviour of the jay *Garrulus glandarius*. *Ibis* **93**:414–442, 602–625.

Gorlick, D. L. (1976). Dominance hierarchies and factors influencing dominance in the guppy *Poecilia reticulata* (Peters). *Anim. Behav.* **24**:336–346.

Gowaty, P. A., and Gauthreaux, S. A., Jr. (1977). The influence of social dominance on mating systems. In preparation.

Grant, P. R. (1970). Experimental studies of competitive interaction in a two-species system. II: The behaviour of *Microtus, Peromyscus* and *Clethrionomys* species. *Anim. Behav.* **18**:411–426.

Grant, T. R. (1973). Dominance and association among members of a captive and a free-ranging group of grey kangaroos (*Macropus giganteus*). *Anim. Behav.* **21**:449–456.

Greenberg, B. (1946). The relation between territory and social hierarchy in the green sunfish. *Anat. Rec.* **94**:395.

Greenwood, P. J., and Harvey, P. H. (1976). The adaptive significance of variation in breeding area fidelity in the blackbird (*Turdus merula* L.). *J. Anim. Ecol.* **45**:887–898.

Gross, A. O. (1940). The migration of Kent Island herring gulls. *Bird-Banding* **11**:129–155.

Grzimek, G. (1949). Rangordnungversuche mit Pferden. *Z. Tierpsychol.* **6**:455–464.

Guhl, A. M. (1953). Social behavior of the domestic fowl. Tech. Bull. 73 (June) Kansas St. Coll., Agr. Exp. Stn., Manhattan, Kan.

Guhl, A. M., and Allee, W. C. (1944). Some measurable effects of social organization in flocks of hens. *Physiol. Zool.* **17**:320–347.

Guhl, A. M., Collias, N. E., and Allee, W. C. (1945). Mating behavior and the social hierarchy in small flocks of white leghorns. *Physiol. Zool.* **18**:365–390.

Hamilton, W. D. (1971). Geometry for the selfish herd. *J. Theoret. Biol.* **31**:295–312.

Haneda, K., and Teranishi, K. (1968). [Life history of the eastern great reed warbler (*Acrocephalus arundinaceus orientalis*). II: Polygyny and territory.] *Jap. J. Ecol.* **18**:204–212.

Hartzler, J. E. (1970). Winter dominance relationship in black-capped chickadees. *Wilson Bull.* **82**:427–434.

Haubrich, R. (1961). Hierarchical behaviour in the South African clawed frog, *Xenopus laevis* Daudin. *Anim. Behav.* **9**:71–76.

Hazlett, B. A. (1966). Social behavior of the Paguridae and Diogenidae of Curacao. *Studies on the Fauna of Curacao and Other Caribbean Islands* (The Hague) **23**:1–143.

Hazlett, B. A. (1970). The effect of shell size and weight on the agonistic behavior of a hermit crab. *Z. Tierpsychol.* **27**:369–374.

Healey, M. C. (1967). Aggression and self-regulation of population size in deermice. *Ecology* **48**:377–392.

Heape, W. (1931). *Emigration, Migration and Nomadism,* W. Heffer & Sons, Cambridge, Mass., 369 pp.

Heldmann, G. (1936a). Über die Entwicklung der polygynen Wabe von *Polistes gallica* L. *Arb. Physiol. Angew. Entomol.* **3**:257–259.

Heldmann, G. (1936b). Über das Leben auf Waben mit mehreren überwinterten Weibchen von *Polistes gallica* L. *Biol. Zentralbl.* **56**:389–401.

Hinde, R. A. (1955). A comparative study of the courtship of certain finches (Fringillidae). *Ibis* **97**:706–745.

Hinde, R. A. (1956). A comparative study of the courtship of certain finches (Fringillidae). *Ibis* **98**:1–23.

Hodge, M. H., Peacock, L. J., and Hoff, L. A. (1967). The effect of age and food deprivation upon the general activity of the rat. *J. Genet. Psychol.* **111**:135–145.

Hoffer, E. (1882). *Die Hummeln Steiermarks: Lebengeschichte und Beschreibung Derselben,* Part 1, Leuschner & Lubensky, Graz, 92 pp.

Höhn, E. O. (1967). Observations on the breeding biology of Wilson's phalarope (*Steganopus tricolor*) in central Alberta. *Auk* **84**:220–244.

Howard, W. E. (1960). Innate and environmental dispersal of individual vertebrates. *Am. Midl. Nat.* **63**:152–161.

Huber, P. (1802). Observations on several species of the genus *Apis,* known by the name of humble-bees, and called Bombinatrices by Linnaeus. *Trans. Linn. Soc. London* **6**:214–298.

Hughes, B. O., and Wood-Gush, D. G. M. (1973). An increase in activity of domestic fowls produced by nutritional deficiency. *Anim. Behav.* **21**:10–17.

Hughes, B. O., Wood-Gush, D. G. M., and Jones, R. M. (1974). Spatial organization in flocks of domestic fowls. *Anim. Behav.* **22**:438–445.

Imboden, C. (1974). Zug, Fremdansiedlung und Brutperiode des Kiebitz *Vanellus vanellus* in Europa. *Orn. Beob.* **71**:5–134.

James, W. T. (1939). Further experiments in social behavior among dogs. *J. Genet. Psychol.* **54**:151–164.

Jenkins, D. W. (1944). Territory as a result of depotism and social organization in geese. *Auk* **61**:30–47.

Jenkins, T. M., Jr. (1969). Social structure, position choice and microdistribution of two trout species (*Salmo trutta* and *Salmo gairdneri*) resident in mountain streams. *Anim. Behav. Monogr.* **2**:55–123.

Jensen, D. D. (1970). Polythetic biopsychology: An alternative to behaviorism. In Reynierse, J. H. (ed.), *Current Issues in Animal Learning,* University of Nebraska Press, Lincoln.

Johnsgard, P. A. (1967). Dawn rendezvous on the lek. *Nat. Hist.* **76** (March):16–21.

Johnston, J. S., and Heed, W. B. (1975). Dispersal of *Drosophila*: The effect of baiting on the behavior and distribution of natural populations. *Am. Nat.* **109**:207–216.

Johnston, R. F. (1956). Population structure in salt marsh song sparrows. Part 1: Environment and annual cycle. *Condor* **5**:24–44.

Jolly, A. (1966). *Lemur Behavior: A Madagascar Field Study,* University of Chicago Press, Chicago, 187 pp.

Jolly, A. (1972). *The Evolution of Primate Behavior,* Macmillan, New York, 397 pp.

Katz, D., and Toll, A. (1923). Die Messung von Charakter- und Begabungsunterschieden bei Tieren. (Versuch mit Hühnern). *Z. Psychol. Physiol. Sinnesorg. Abt. I* **93**:287–311.

Ketterson, E. D., and Nolan, V. (1976). Geographic variation and its climatic correlates in the sex ratio of eastern-wintering dark-eyed juncos (*Junco hyemalis hyemalis*). *Ecology* 57:679–693.

Kikkawa, J. (1961). Social behaviour of the white-eye *Zosterops lateralis* in winter flocks. *Ibis* 103:428–442.

King, J. A. (1955). Social behavior, social organization, and population dynamics in a black-tailed prairiedog town in the Black Hills of South Dakota. *Contrib. Lab. Vertebr. Biol. Univ. Mich.*, No. 67:1–123.

Kinsey, K. P. (1976). Social behaviour in confined populations of the Allegheny woodrat, *Neotoma floridana magister*. *Anim. Behav.* 24:181–187.

Kistchinski, A. A. (1975). Breeding biology and behaviour of the grey phalarope *Phalaropus fulicarius* in east Siberia. *Ibis* 117:285–301.

Klopfer, P. H., and Klopfer, M. S. (1973). How come leaders to their posts? The determination of social ranks and roles. *Am. Sci.* 61:560–564.

Kluyver, H. N. (1957). Roosting habits, sexual dominance and survival in the great tit. *Cold Spring Harbor Symp. Quant. Biol.* 22:281–285.

Kluyver, H. N., and Tinbergen, L. (1953). Territory and regulation of density in titmice. *Arch. Neerl. Zool.* 10:265–286.

Krebs, C. J., Gaines, M. S., Keller, B. L., Myers, J. H., and Tamarin, R. H. (1973). Population cycles in small rodents. *Science* 179:35–41.

Kroh, O. (1927). Weitere Beiträge zur Psychologie des Haushuhns. *Z. Psychol.* 103:203–227.

Kruuk, H. (1972). *The Spotted Hyena: A Study of Predation and Social Behavior*, University of Chicago Press, Chicago, 335 pp.

Lack, D. (1943). *The Life of the Robin*, H. F. & G. Witherby, London, 224 pp.

Lack, D. (1944). The problem of partial migration. *Br. Birds* 37:122–130, 143–150.

Lack, D. (1954). *The Natural Regulation of Animal Numbers*, Oxford University Press, London, 343 pp.

Lack, D. (1966). *Population Studies of Birds*, Clarendon Press, Oxford, 341 pp.

Lack, D. (1968). Bird migration and natural selection. *Oikos* 19:1–9.

LaFollette, R. M. (1971). Agonistic behavior and dominance in confined wallabies, *Wallabia rufogrisea frutica*. *Anim. Behav.* 19:93–101.

Landau, H. G. (1951a). On dominance relations and the structure of animal societies. I: Effect of inherent characteristics. *Bull. Math. Biophys.* 13:1–19.

Landau, H. G. (1951b). On dominance relations and the structure of animal societies. II: Some effects of possible social factors. *Bull. Math. Biophys.* 13:245–262.

Landau, H. G. (1965). Development of structure in a society with a dominance relation when new members are added successively. *Bull. Math. Biophys.* 27:151–160.

Landau, H. G. (1968). Models of social structure. *Bull. Math. Biophys.* 30:215–224.

Lange, R. (1967). Die Nahrungsverteilung unter den Arbeiterinnen des Waldameisenstaates. *Z. Tierpsychol.* 24:513–545.

Lawlor, M. (1963). Social dominance in the golden hamster. *Bull. Br. Psychol. Soc.* 16:25–38.

LeBoeuf, B. J. (1972). Sexual behavior in the northern elephant seal *Mirounga angustirostris*. *Behaviour* 41:1–26.

LeBoeuf, B. J., and Peterson, R. S. (1969). Social status and mating activity in elephant seals. *Science* 163:91–93.

Levins, R. (1970a). Extinction. In *Some Mathematical Questions in Biology*, Vol. 2, American Mathematical Society, Providence, R.I., pp. 75–107.

Levins, R. (1970b). Fitness and optimization. In Koyima, K. (ed.), *Mathematical Topics in Population Genetics*, Springer, New York, pp. 389–400.

Leyhausen, P. (1956). Verhaltensstudien an Katzen. *Z. Tierpsychol. Suppl.* 2, 120 pp.

Leyhausen, P. (1971). Dominance and territoriality as complemented in mammalian social structure. In Essler, A. H. (ed.), *Behavior and Environment: The Use of Space by Animals and Men,* Plenum, New York, pp. 22–33.

Lidicker, W. Z., Jr. (1962). Emigration as a possible mechanism permitting the regulation of population density below carrying capacity. *Am. Nat.* **96**:29–33.

Lidicker, W. Z., Jr. (1975). The role of dispersal in the demography of small mammals. In Golley, F. B., Petrusewicz, K., and Ryszkowski, L. (eds.), *Small Mammals: Their Productivity and Population Dynamics,* Cambridge University Press, Cambridge, pp. 103–128.

Lockie, J. (1956). Winter fighting in feeding flocks of rooks, jackdaws, and carrion crows. *Bird Study* **3**:180–190.

Löhrl, H. (1959). Zur Frage des Zeitpunktes einer Prägung auf die Heimatregion beim Halsbandschnäpper (*Ficedula albicollis*). *J. Ornithol.* **100**:132–140.

Lorenz, K. (1931). Beiträge zur Ethologie sozialer Corviden. *J. Ornithol.* **79**:76–127.

Lorenz, K. (1935). Der Kumpan in der Umwelt des Vogels. *J. Ornithol.* **83**:137–213.

Lowe, M. E. (1956). Dominance-subordinance relationships in the crawfish *Cambarellus shufeldti. Tulane Stud. Zool. Bot.* **4**:139–170.

MacFarland, C. (1972). Goliaths of the Galapagos. *National Geographic* **142**(November):633–649.

MacMillan, R. E. (1964). Population ecology, water relations, and social behavior of a southern California semidesert rodent fauna. *Univ. Calif. Publ. Zool.* **71,** 59 pp.

Marler, P. R. (1956). Behaviour of the chaffinch, *Fringilla coelebs. Behaviour,* suppl. **5,** 184 pp.

Maslow, A. H. (1935). The social behavior of monkeys and apes. *Int. J. Individ. Psychol.* **1**:47–59.

Maslow, A. H. (1936). The role of dominance in the social and sexual behavior of infra-human primates. IV: The determination of hierarchy in pairs and in a group. *J. Genet. Psychol.* **49**:161–198.

Maslow, A. H. (1940). Dominance quality and social behavior in infrahuman primates. *J. Soc. Psychol.* **11**:313–324.

Masure, R. H., and Allee, W. C. (1934). Flock organizations of the shell parakeet, *Melopsittacus undulatus* Shaw. *Ecology* **15**:388–398.

Maynard Smith, J. (1964). Group selection and kin selection. *Nature* **201**:1145–1147.

Maynard Smith, J. (1974). *Models in Ecology,* Cambridge University Press, Cambridge, Mass., 146 pp.

McBride, G. (1963). The "teat order" and communication in young pigs. *Anim. Behav.* **11**:53–56.

McDonald, A. L., Heimstra, N. W., and Damkot, D. K. (1968). Social modification of agonistic behaviour in fish. *Anim. Behav.* **16**:437–441.

McHugh, T. (1958). Social behavior of the American buffalo (*Bison bison bison*). *Zoologica* **43**:1–40.

McKay, F. E. (1971). Behavioural aspects of population dynamics in unisexual-bisexual *Poeciliopsis* (Pisces: Poeciliidae). *Ecology* **52**:778–790.

Meese, G. B., and Ewbank, R. (1973). The establishment and nature of the dominance hierarchy in the domesticated pig. *Anim. Behav.* **21**:326–334.

Mewaldt, L. R. (1976). Winter philopatry in white-crowned sparrows (*Zonotrichia leucophrys*). *N. Am. Bird Bander* **1**:14–20.

Michener, G. R. (1973). Intraspecific aggression and social organization in ground squirrels. *J. Mammal.* **54**:1001–1003.

Michener, H., and Michener, J. R. (1935). Mockingbirds, their territories and individualities. *Condor* **37**:97–140.

Miller, R. S. (1967). Pattern and process in competition. In Cragg, J. B. (ed.), *Advances in Ecological Research,* Vol. 4, Academic Press, London, pp. 1–74.

Miller, R. S., and Stephen, W. J. D. (1966). Spatial relationships in flocks of sandhill cranes (*Grus canadensis*). *Ecology* **47**:323–327.

Minock, M. E. (1971). Social relationships among mountain chickadees (*Parus gambeli*). *Condor* **73**:118–120.

Montagner, H. (1966). Le mécanisme et les conséquences des comportements trophallactiques chez les guêpes du genre *Vespa*. Thesis, Faculté des Sciences de l'Université de Nancy, France, 143 pp.

Moore, W. S., and McKay, F. E. (1971). Coexistence in unisexual–bisexual species complexes of *Poeciliopsis* (Pisces: Poeciliidae). *Ecology* **52**:791–799.

Morimoto, R. (1961a). On the dominance order in *Polistes* wasps. I: Studies on the social Hymenoptera in Japan XII. *Sci. Bull. Facul. Agr., Kyushu Univ.* **18**:339–351.

Morimoto, R. (1961b). On the dominance order in *Polistes* wasps. II: Studies on the social Hymenoptera in Japan XIII. *Sci. Bull. Facul. Agr., Kyushu Univ.* **19**:1–17.

Morris, G. K. (1971). Aggression in male conocephaline grasshoppers (Tettigoniidae). *Anim. Behav.* **19**:132–137.

Morse, D. H. (1974). Niche breadth as a function of social dominance. *Am. Nat.* **109**:818–830.

Moyer, J. T. (1976). Geographical variation and social dominance in Japanese populations of the anemonefish *Amphiprion clarkii. Jap. J. Ichthyol.* **23**:12–22.

Murchison, C. (1935a). The experimental measurement of a social hierarchy in *Gallus domesticus*. II: The identification and inferential measurement of social reflex no. 1, and social reflex no. 2 by means of social discrimination. *J. Soc. Psychol.* **6**:3–30.

Murchison, C. (1935b). The experimental measurement of a social hierarchy in *Gallus domesticus*. III: The direct and inferential measurement of social reflex no. 3. *J. Genet. Psychol.* **46**:76–102.

Murchison, C. (1935c). The experimental measurement of a social hierarchy in *Gallus domesticus*. IV: Loss of body weight under conditions of mild starvation as a function of social dominance. *J. Gen. Psychol.* **12**:296–311.

Murchison, C. (1935d). The experimental measurement of a social hierarchy in *Gallus domesticus*. VI: Preliminary identification of social law. *J. Gen. Psychol.* **12**:227–248.

Murchison, C. (1936). The time function in the experimental formation of social hierarchies of different sizes in *Gallus domesticus. J. Soc. Psychol.* **7**:3–18.

Murchison, C., Pomerat, C. M., and Zarrow, M. X. (1935). The experimental measure of a social hierarchy in *Gallus domesticus*. V: The post-mortem measurement of anatomical features. *J. Soc. Psychol.* **6**:172–181.

Murray, B. G., Jr. (1967). Dispersal in vertebrates. *Ecology* **48**:975–978.

Murton, R. K. (1968). Some predator–prey relationships in bird damage and population control. In Murton, R. K., and Wright, E. N. (eds.), *The Problems of Birds as Pests*, Symp. Inst. Biol. **17**:157–169.

Murton, R. K., Isaacson, A. J., and Westwood, N. J. (1966). The relationships between woodpigeons and their clover food supply and the mechanism of population control. *J. Appl. Ecol.* **3**:55–96.

Myers, J. H., and Krebs, C. J. (1971). Genetic, behavioral, and reproductive attributes of dispersing field voles *Microtus pennsylvanicus* and *Microtus ochrogaster. Ecol. Monogr.* **41**:53–78.

Myers, K., Hale, C. S., Mykytowycz, R. and Hughes, R. L. (1971). The effects of varying density and space on sociality and health in animals. In Esser, A. H. (ed.), *Behavior and Environment: The Use of Space by Animals and Men*, Plenum, New York, pp. 148–187.

Mykytowycz, R. (1962). Territorial function of chin gland secretion in the rabbit, *Oryctolagus cuniculus* (L.) *Nature* **193**:799.

Myrberg, A. A. (1972). Social dominance and territoriality in the bicolor damselfish. *Behaviour* **51**:207–231.

Narise, T. (1968). Migration and competition in *Drosophila*. I: Competition between wild and vestigial strains of *Drosophila melanogaster* in a cage and migration-tube population. *Evolution* 22:301–306.

Nelson, B. A., and Janson, R. G. (1949). Starvation of pheasants in South Dakota. *J. Wildl. Manage.* 13:308–309.

Newton, I. (1970). Irruptions of crossbills in Europe. In Watson, A. (ed.), *Animal Populations in Relation to Their Food Resources*, Symp. Brit. Ecol. Soc., No. 10, Blackwell Scientific Publishers, Oxford, pp. 337–353.

Nice, M. M. (1937). Studies in the life history of the song sparrow, I. *Trans. Linn. Soc. New York* 4:i–vi, 1–247.

Nicolai, J. (1956). On the biology and ethology of the bullfinch (*Pyrrhula pyrrhula* L.). *Z. Tierpsychol.* 18:93–132.

Nisbet, I. C. T., and Medway, L. (1972). Dispersion, population ecology and migration of eastern great reed warblers *Acrocephalus orientalis* wintering in Malyasia. *Ibis* 114:451–494.

Noble, G. K. (1939). The role of dominance in the social life of birds. *Auk* 56:263–273.

Noble, G. K., and Wurm, M. (1943). Social behavior of the laughing gull. *Ann. N.Y. Acad. Sci.* 45:179–220.

Noble, G. K., Wurm, M., and Schmidt, A. (1938). Social behavior of the black-crowned night heron. *Auk* 55:7–40.

Oakeshott, J. G. (1974). Social dominance, aggressiveness and mating success among male house mice (*Mus musculus*). *Oecologia* 15:143–158.

Odum, E. P. (1942). Annual cycle of the black-capped chickadee, 3. *Auk* 59:499–531.

O'Shea, T. J. (1976). Home range, social behavior, and dominance relationships in the African striped ground squirrel, *Xerus rutilus*. *J. Mammal.* 57:450–460.

Pack, J. C., Mosby, H. S., and Siegel, P. B. (1967). Influence of social hierarchy on gray squirrel behavior. *J. Wildl. Manage.* 31:720–728.

Pardi, L. (1940). Ricerche sui Polistini. I: Poliginia vera ed apparente in *Polistes gallicus* (L.). *Proc. Verb. Soc. Toscana Sci. Nat. Pisa* 49:3–9.

Pardi, L. (1948). Dominance order in *Polistes* wasps. *Physiol. Zool.* 21:1–13.

Parsons, P. A. (1963). Migration as a factor in natural selection. *Genetics* 33:184–206.

Penny, R. L. (1968). Territorial and social behaviour in the Adelie penguin. *Antarct. Res. Ser.* 12:83–131.

Perdeck, A. C. (1958). Two types of orientation in migrating starlings, *Sturnus vulgaris* L. and chaffinches, *Fringilla coelebs* L. as revealed by displacement experiments. *Ardea* 46:1–37.

Perrins, A. C. (1963). Survival in the great tit, *Parus major*. *Proc. XIII Internat. Ornithol. Congr.*, pp. 717–728.

Peterson, R. S., and Bartholomew, G. A. (1967). *The Natural History and Behavior of the California Sea Lion*, American Society of Mammalogists, Stillwater, Okla., 79 pp.

Picozzi, N., and Weir, D. (1976). Dispersal and causes of death of buzzards. *Br. Birds* 69:193–201.

Pilters, H. (1954). Untersuchungen über angeborene Verhaltensweisen bei Tylopoden, unter besonderer Berücksichtigung der neuweltlichen Formen. *Z. Tierpsychol.* 11:213–303.

Pinowski, J. (1965). Overcrowding as one of the causes of dispersal of young tree sparrows. *Bird Study* 12:27–33.

Poirier, F. E. (1970). Dominance structure of the Nilgiri langur (*Presbytis johnii*) of south India. *Folia Primatol.* 12:161–186.

Poole, T. B., and Morgan, H. D. R. (1976). Social and territorial behaviour of laboratory mice (*Mus musculus* L.) in small complex areas. *Anim. Behav.* 24:476–480.

Price, E. O., Belanger, P. L., and Duncan, R. A. (1976). Competitive dominance of wild and domestic Norway rats (*Rattus norvegicus*). *Anim. Behav.* **24**:589–599.

Ralls, K. (1976). Mammals in which females are larger than males. *Q. Rev. Biol.* **51**:245–276.

Ralph, C. J., and Mewaldt, L. R. (1975). Timing of site fixation upon the wintering grounds in sparrows. *Auk* **92**:698–705.

Rashevsky, N. (1959). *Mathematical Biology of Social Behavior,* University of Chicago Press, Chicago, 256 pp.

Raveling, D. G. (1970). Dominance relationships and agonistic behavior of Canada geese in winter. *Behaviour* **37**:291–319.

Reimer, J. D., and Petras, M. L. (1967). Breeding structure of the house mouse, *Mus musculus*, in a population cage. *J. Mammal.* **48**:88–89.

Reynierse, J. H., Manning, A., and Cafferty, D. (1972). The effects of hunger and thirst on body weight and activity in the cockroach (*Nauphoeta cinerea*). *Anim. Behav.* **20**:751–757.

Rheinwald, G., and Gutscher, H. (1969). Dispersion und Ortstreue der Mehlschwalbe (*Delichon urbica*). *Vogelwelt* **90**:121–140.

Richards, S. M. (1974). The concept of dominance and methods of assessment. *Anim. Behav.* **22**:914–930.

Roberts, P. (1971). Social interactions of *Galago crassicaudatus*. *Folia Primatol.* **14**:171–181.

Roff, D. (1975). Population stability and the evolution of dispersal in a heterogeneous environment. *Oecologia* **19**:217–237.

Rovner, J. S. (1968). Territoriality in the sheet-web spider *Linyphia triangularis* (Clerck) (Araneae, Linyphiidae). *Z. Tierpsychol.* **25**:232–242.

Rowell, T. E. (1963). The social development of some rhesus monkeys. In Foss, B. (ed.), *Determinants of Infant Behavior,* Methuen, London, pp. 35–44.

Rowell, T. E. (1971). Organization of caged groups of *Cercopithecus* monkeys. *Anim. Behav.* **19**:625–645.

Rowell, T. E. (1974). The concept of social dominance. *Behav. Biol.* **11**:131–154.

Rowley, I. (1965). The life history of the superb blue wren *Malurus cyaneus*. *Emu* **64**:251–297.

Rowley, M. H., and Christian, J. J. (1976). Intraspecific aggression of *Peromyscus leucopus*. *Behav. Biol.* **17**:249–253.

Russell, E. M. (1970). Agonistic interactions in the red kangaroo (*Megaleia rufa*). *J. Mammal.* **51**:80–88.

Sabine, W. S. (1949). Dominance in winter flocks of juncos and tree sparrows. *Physiol. Zool.* **22**:68–85.

Sabine, W. S. (1959). The winter society of the Oregon junco: Intolerance, dominance, and the pecking order. *Condor* **61**:110–135.

Sade, D. S. (1967). Determinants of dominance in a group of free-ranging rhesus monkeys. In Altmann, S. A. (ed.), *Social Communication Among Primates,* University of Chicago Press, Chicago, pp. 99–114.

Sadleir, R. M. F. S. (1965). The relationship between agonistic behaviour and population changes in the deermouse, *Peromyscus maniculatus* (Wagner). *J. Anim. Ecol.* **34**:331–352.

Sadleir, R. M. F. S. (1970). The establishment of a dominance rank order in male *Peromyscus maniculatus* and its stability with time. *Anim. Behav.* **18**:55–59.

Sakagami, S. F. (1954). Occurrence of an aggressive behaviour in queenless hives, with considerations on the social organization of honeybees. *Insectes Sociaux* **1**:331–343.

Sakagami, S. F., Montenegro, M. J., and Kerr, W. E. (1965). Behavior studies of the stingless bees, with special reference to the oviposition process: 5. *Melipona quadrifasciata anthidiodes* Lepeletier. *J. Facul. Sci., Hokkaido University,* 6th series (Zoology) **15**:578–607.

Schaller, G. B. (1967). *The Deer and the Tiger: A Study of Wildlife in India,* University of Chicago Press, Chicago, 370 pp.

Schein, M. W., ed. (1975). *Social Hierarchy and Dominance.* Benchmark Papers in Animal Behavior, Vol. 3, Halsted Press, New York, 401 pp.

Schein, M. W., and Fohrman, M. H. (1955). Social dominance relationships in a herd of dairy cattle. *Br. J. Anim. Behav.* **3**:45–55.

Schenkel, R. (1947). Ausdrucks-Studien an Wölfen. Gefangenschafts-Beobachtungen. *Behaviour* **1**:81–129.

Scheven, J. (1958). Beitrag zur Biologie der Schmarotzerfeldwespen *Sulcopolistes atrimandibularis* Zimm., *S. semenowi* F. Morawitz und *S. sulcifer* Zimm. *Insectes Sociaux* **5**:409–437.

Schjelderup-Ebbe, T. (1922a). Beiträge zur Sozialpsychologie des Haushuhns. *Z. Psychol.* **88**:225–252.

Schjelderup-Ebbe, T. (1922b). Soziale Verhältnisse bei Vögeln. *Z. Psychol.* **90**:106–107.

Schjelderup-Ebbe, T. (1923). Weitere Beiträge zur Sozial- und Individual-psychologie des Haushuhns. *Z. Psychol.* **92**:60–87.

Schjelderup-Ebbe, T. (1931). Die Despotie im sozialen Leben der Vögel. In *Arbeiten zur Biologischen Grundlegung der Soziologie,* Leipzig, pp. 77–137.

Schloeth, R. (1958). Cycle annuel et comportement social du taureau de Camargue. *Mammalia* **22**:121–139.

Schwartz, P. (1963). Orientation experiments with northern waterthrushes wintering in Venezuela. *Proc. XIII Internat. Ornithol. Congr.,* pp. 481–484.

Scott, J. P. (1945). Social behavior, organization and leadership in a small flock of domestic sheep. *Comp. Psychol. Monogr.,* Serial 96, **18**:1–29.

Scott, J. P. (1946). Dominance reactions in a small flock of goats. *Anat. Rec.* **94**:380–390.

Shaw, M. J. P. (1967). The flight behaviour of alate *Aphis fabae. Rep. Rothamsted Exp. Stn.* (1966):200–201.

Shaw, M. J. P. (1968). The flight behaviour and wing polymorphism of alate *Aphis fabae. Rep. Rothamsted Exp. Stn.* (1967):191–193.

Sheppe, W. A. (1966). Social behavior of the deermouse *Peromyscus leucopus* in the laboratory. *Wasmann J. Biol.* **24**:49–65.

Shoemaker, H. H. (1939). Social hierarchy in flocks of the canary. *Auk* **56**:381–406.

Simmons, K. E. L. (1970). Ecological determinants of breeding adaptations and social behaviour in two fish-eating birds. In Crook, J. H. (ed.), *Social Behaviour in Birds and Mammals,* Academic Press, New York, pp. 37–77.

Sladen, W. J. L. (1958). The pygoscelid penguins. I: Methods of Study; II: The Adelie penguin *Pygoscelis adeliae* (Hombron & Jacquinot). *Falkland Isl. Depend. Surv. Sci. Rep.* **17**:1–97.

Smith, S. M. (1976). Ecological aspects of dominance hierarchies in black-capped chickadees. *Auk* **93**:95–107.

Snyder, N. F. R., and Wiley, J. W. (1976). Sexual size dimorphism in hawks and owls of North America. *Ornithol. Monogr.* no. 20, 96 pp.

Sorenson, M. W. (1974). A review of aggressive behavior in the tree shrews. In Holloway, R. L. (ed.), *Primate Aggression, Territoriality, and Xenophobia,* Academic Press, New York, pp. 13–30.

Southwick, C. H., and Siddiqi, M. R. (1967). The role of social tradition in the maintenance of dominance in wild rhesus group. *Primates* **8**:341–353.

Spurr, E. B. (1972). Social organisation of the Adelie penguin, *Pygoscelis adeliae.* Ph.D. thesis, University of Canterbury, Christchurch, New Zealand, 169 pp.

Spurr, E. B. (1974). Individual differences in aggressiveness of Adelie penguins. *Anim. Behav.* **22**:611–616.

Sussman, R. W., and Richard, A. (1974). The role of aggression among diurnal prosimians. In Holloway, R. L. (ed.), *Primate Aggression, Territoriality, and Xenophobia,* Academic Press, New York, pp. 49–76.

Syme, G. J. (1974). Competitive orders as measures of social dominance. *Anim. Behav.* **22**:931–940.

Syme, G. J., Pollard, J. S., Syme, L. A., and Reid, R. M. (1974). An analysis of the limited access measure of social dominance in rats. *Anim. Behav.* **22**:486–500.

Thiessen, D. D. (1964). Population density and behavior: A review of theoretical and physiological contributions. *Texas Rep. Biol. Med.* **22**:266–314.

Thines, G., and Heuts, B. (1968). The effect of submissive experience on dominance and aggressive behavior of *Xiphophorus* (Pisces, Poeciliidae). *Z. Tierpsychol.* **25**:139–154.

Thompson, W. L. (1960). Agonistic behavior in the house finch. 2: Factors in aggressiveness and sociality. *Condor* **62**:378–402.

Tinbergen, N. (1939). Field observations of east Greenland birds. II: The behavior of the snow bunting (*Plectrophenax nivalis subnivalis* [Brehm]) in spring. *Trans. Linn. Soc. N.Y.* **5**:1–94.

Tordoff, H. B. (1954). Social organization and behavior in a flock of captive, nonbreeding red crossbills. *Condor* **56**:346–358.

Uhrich, J. (1938). The social hierarchy in albino mice. *J. Comp. Psychol.* **25**:373–413.

Ulfstrand, S. (1963). Ecological aspects of irruptive bird migration in northwestern Europe. *Proc. XIII Internat. Ornithol. Congr.,* pp. 780–794.

Uvarov, B. P. (1961). Quantity and quality in insect populations. *Proc. Roy. Ent. Soc. London,* Series C, **25**:52–59.

Valverde, J. A. (1959). Moyens d'expression et hiérarchie social chez le vautour fauve *Gyps fulvus* (Habliz). *Alauda* **27**:1–15.

Van Valen, L. (1971). Group selection and the evolution of dispersal. *Evolution* **25**:591–598.

Vessey, S. H. (1971). Free-ranging rhesus monkeys: Behavioural effects of removal, separation and reintroduction of group members. *Behaviour* **40**:216–227.

Vos, A. de, Brokx, P., and Geist, V. (1967). A review of social behavior of the North American cervids during the reproductive period. *Am. Midl. Nat.* **77**:390–417.

Wald, G., and Jackson, B. (1944). Activity and nutritional deprivation. *Proc. Natl. Acad. Sci. U.S.A.* **30**:255–263.

Waloff, N. (1973). Dispersal by flight of leafhoppers (Auchenorrhyncha: Homoptera). *J. Appl. Ecol.* **10**:705–730.

Ward, P. (1965). Feeding ecology of the black-faced dioch *Quelea quelea* in Nigeria. *Ibis* **107**:173–214.

Watson, A., and Moss, R. (1970). Dominance, spacing behaviour and aggression in relation to population limitation in vertebrates. In Watson, A. (ed.), *Animal Populations in Relation to Their Food Resources,* Blackwell Scientific Publishers, Oxford, pp. 167–218.

Watson, J. R. (1970). Dominance–subordination in caged groups of house sparrows. *Wilson Bull.* **82**:268–278.

Watts, C. R., and Stokes, A. W. (1971). The social order of turkeys. *Sci. Am.* **224** (June):112–118.

Wellington, W. G. (1964). Qualitative changes in population in unstable environments. *Can. Entomol.* **96**:436–451.

Westby, G. W. M. (1974). Further analysis of the individual discharge characteristics predicting social dominance in the electric fish, *Gymnotus carapo. Anim. Behav.* **23**:249–260.

Williams, W. T., Kikkawa, J., and Morris, D. K. (1972). A numerical study of agonistic behaviour in the grey-breasted silvereye (*Zosterops lateralis*). *Anim. Behav.* **20**:155–165.

Willis, E. O. (1967). The behavior of bicolored antbirds. *Univ. Calif. Publ. Zool.* **79**:1–127.

Willis, E. O. (1968). Studies of lunulated and Salvin's antbirds. *Condor* **70**:128–148.

Wilson, A. (1968). Social behaviour of free-ranging rhesus monkeys with an emphasis on aggression. Ph.D. thesis, University of California, Berkeley, 143 pp.

Wilson, E. O. (1971). *The Insect Societies,* Belknap Press, Cambridge, Mass., 548 pp.

Wilson, E. O. (1975). *Sociobiology,* Harvard University Press, Cambridge, Mass., 697 pp.

Wood-Gush, D. G. M. (1971). *The Behaviour of the Domestic Fowl,* Heinemann, London, 147 pp.

Wynne-Edwards, V. C. (1962). *Animal Dispersion in Relation to Social Behavior,* Hafner Co., New York, 653 pp.

Yarnall, J. L. (1969). Aspects of the behaviour of *Octopus cyanea* Gray. *Anim. Behav.* **17**:747–754.

Yoshikawa, K. (1963). Introductory studies on the life economy of polistine wasps. 2: Super-individual stage; 3: Dominance order and territory. *J. Biol., Osaka City Univ.* **14**:55–61.

Zahavi, A. (1974). Communal nesting in the Arabian babbler. *Ibis* **116**:84–87.

Chapter 3

HOW DOES BEHAVIOR DEVELOP?

P. P. G. Bateson

Sub-Department of Animal Behaviour
University of Cambridge
Madingley, Cambridge CB3 8AA, England, U.K.

I. ABSTRACT

The developmental origins of behavior are undoubtedly multiple and variegated in action. It is essential to formulate theories of how the major sources of variation in behavior interact to generate observed outcomes. General principles will doubtless grow out of specific instances. I argue that the most likely way in which they will do so is if competing explanations for particular phenomena are explicitly set in opposition to each other. In order to illustrate how this might happen, I consider a variety of different explanations for apparent discontinuities in behavioral development—phenomena that are exciting considerable interest at the moment. Some of the explanations simply point to deficiencies of method; some suggest that after a loss of the original rank order on a particular measure, that order is recovered; and some suggest ways in which a change in rank order becomes permanent.

II. INTRODUCTION

Social behavior can be examined from many different standpoints that are not always recognized as being distinct. At one time, the major preoccupation was with short-term control. In the last few years, the fashions of sociobiology have focused attention on the function of social behavior and its evolutionary history. The emphasis on adaptiveness immediately raised a different set of questions about the origins of social behavior. It does not follow, of course, that because a particular character helps animals to survive and to propagate that character from one generation to the next, the method of transmission is exclusively genetic. Examples of animals' acquiring the detailed species-specific characteristic of the behavior by learning are commonplace. So the evolutionary and functional questions have raised interest once again in the entirely different problem of behavioral development.

Fortunately, analysis of the development origins of behavior is currently being conducted with considerably greater precision and clarity of thought than was the case even a few years ago. The conditions necessary for the development of a behavior pattern are now thought to act in a number of quite different ways. They may initiate developmental processes, they may facilitate processes that are already under way, or they may maintain the end products in the behavioral repertoire. Gottlieb (1976) in particular has argued for these admittedly arbitrary distinctions in relation to "experience," but they can be applied with equal force to the inherited factors that are necessary for the development of a given behavior pattern. In addition, the consequences of developmental determinants are recognized as ranging from the highly specific to the general (Lehrman, 1970; Bateson, 1976b). Just where a line could or should be drawn is decidedly a matter of opinion, but the distinction between specific and general effects does at least provide the basis for a preliminary classification.

The product of a great deal of thought and debate can be distilled into a matrix that should be usable for all the conditions that are necessary for the development of a given pattern of behavior (see Table I). Any developmental determinant should lie somewhere in that matrix. In as much as variation in that determinant is a major source of variation in behavior, it must be identified if development is to be adequately analyzed and understood. It seems reasonably likely that many patterns of behavior of an adult human being are determined by a *combination* of conditions from all *12* categories, which is a daunting thought. If correct, this means that the views of development implicit in most studies are far too simplistic. The process of development has been thought of as one in which a straightforward relationship exists between the starting points and the end points. For instance, it was a goal of Konrad Lorenz's school of ethology to analyze behavior into genetically determined and environmentally determined components that were supposedly intercalated in the adult behavior (e.g. Lorenz, 1965; Eibl-Eibesfeldt, 1970). Moreover much of the experimental work on behavioral development has consisted of manipulating the genotype or some aspect of early rearing conditions and looking for a difference between the groups when adult. Similarly, clinicians have measured, say, a facet of birth trauma in human babies and hunted for associations between that and behavior shown later in life.

Table I. A Classification of All the Developmental Determinants That May Be Necessary for the Distinctiveness of a Given Pattern of Behavior

		Specific consequences	General consequences
Inherited determinants	Initiating	1	4
	Facilitating	2	5
	Maintaining	3	6
Environmental determinants	Initiating	7	10
	Facilitating	8	11
	Maintaining	9	12

Nobody should denigrate the immense amount of work put into such developmental studies nor belittle the positive achievements. Furthermore, while the approaches may appear unsubtle to modern eyes, they were frequently seen as the only way to begin tackling an immensely difficult task. Some of the studies set out to show that particular types of factors, such as potentially relevant experience, were *not* important. Nevertheless, a great deal of the negative evidence, and there is a lot of it, was unwanted and pointed to some serious shortcomings in the theoretical frameworks. The shortcomings spring in large part from the notion that each distinctive characteristic of adult behavior is determined by only a few inherited factors and/or only a few environmental ones. Furthermore, it was thought that these developmental determinants acted in a straightforward chainlike manner.

Gradually an awareness is growing that what is needed is an approach that will cope with (1) the multiple and variegated nature of developmental determination and (2) the interactions that take place between the determinants. While this amounts to a systems approach, it does not mean that a ready-made theory can be plucked off some shelf and instantly applied to a given aspect of behavioral development. The theories must arise from particular cases.

Partial theories of behavioral development do, of course, exist (e.g., Schneirla, 1965; Scott *et al.* 1974; Bateson, 1976a; Isaac and O'Connor, 1976). But they generally meet with considerable sales resistance. Disinterested onlookers who go to the bother of understanding what is being proposed are commonly troubled by the *ad hoc* character of the theories and the limbo in which they appear to float. Why *that* particular assumption? Why select *this* evidence? And, perhaps most important of all, in how many *other* ways could the data be explained? It is not so much that the theories are untestable but that the outsider is left wondering what is at

stake. Why go to the bother of testing the predictions? In part, the value of
the ideas may be in doubt just because they have not yet been put to work.
But it is often because the alternative explanations are not stated and there
is no comparative knowledge on which to base a judgment about the
theorist's particular preference. For instance, Scott *et al.* (1974) referred to
the theory of critical periods when, as far as one can judge, the evidence
that they were attempting to explain could conceivably arise from half a
dozen different kinds of process. It is not at all clear why Scott *et al.* preferred their version and why it should be taken especially seriously. I should
hasten to add that exactly comparable criticisms can be leveled at some of
my own theorizing. In the defense of the theorist, commitment to generating *one* working model is hard enough, let alone half a dozen. Nevertheless,
I think that it is important for the health of the subject that when a
phenomenon is to be explained, as many alternative explanations as possible
are set out in opposition to each other. When that is done, judgments about
the worth of a particular set of ideas become much easier and the problems
for analysis are likely to be much more clearly formulated.

In order to develop the argument, I want to take a developmental
phenomenon and consider a variety of different ways in which it might be
explained. (I shall *not* consider alternative explanations for critical or sensitive periods here since I plan to do that elsewhere.) I shall focus on a
phenomenon that is raised by Kagan elsewhere in this volume, namely, discontinuities in development.

III. DISCONTINUITIES IN DEVELOPMENT

The most direct evidence for a discontinuous change in development
comes from descriptive studies in which it is observed that a qualitatively
new pattern of behavior is given at a certain stage in the growth of the individual. The behavior may consist of something relatively straightforward,
like grooming, or it may be integrated series of actions that imply the
development of a cognitive strategy, such as Piaget's "object permanence."
The taxonomic problems of deciding when a "new" pattern of behavior
really is new for that individual are not trivial. For instance, may they not
be fundamentally equivalent to something that went beforehand? This question is sidestepped by a quite different approach to discontinuity, namely, to
look for changes in rates of change. For example, a growth spurt before
puberty is sometimes regarded as a discontinuity in development. The
assumption is, of course, that the intervals of the same value at different
parts of the scale of measurement are equivalent. This assumption may be

reasonable for height, but it is much more questionable for most indices of behavior. In any event, few patterns of behavior can be easily measured in the same way at widely different stages of development. Because of these difficulties, evidence for discontinuities in behavioral development tends to come from a different quarter.

A discontinuity is often presumed when a given event exerts an influence at one stage in development but subsequently its effects cannot be detected. For example, Dunn (1976) quoted several instances in which abnormally treated children show differences in behavior from normally treated children at one stage of their development but no differences can be obtained when their behavior is sampled later in life (see also Clarke and Clarke, 1976). Another type of evidence for discontinuities comes from longitudinal correlation studies in which the same pattern of behavior is measured repeatedly in a group of subjects through development. Behavior at one stage may successfully predict behavior at the next and the next but not the one after that (see Fig. 1). Kagan has given specific examples in his chapter in this volume. So long as the rank order is dependent on new sources of variation or fluctuates randomly, extreme values tend to regress toward the mean. As a result, correlations between behavior at one stage and behavior at succeeding stages tend to diminish with the elapse of time. Nevertheless, a particularly rapid reduction in the strength of a correlation such as is shown in Fig. 1 requires additional explanation.

Before zooming into the stratosphere with elaborate explanations for what is happening, it is necessary to take a hard and critical look at the

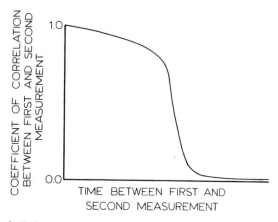

Fig. 1. A hypothetical set of correlations between a measure taken at one stage of development and measures taken at successive stages. A discontinuity in development might be inferred if the values of the correlation suddenly dropped at a given stage, as they do here. Note, however, the need for clear criteria for a "sudden drop."

data. The explanation for a "discontinuity" may be rather trivial. And if it is not, a variety of more interesting interpretations offer themselves. I shall deal with seven possibilities and consider how it may be possible to distinguish between them.

IV. EXPLANATIONS FOR DISCONTINUITIES

A. Reduction in Variability

Positive correlations between a measure taken at one stage of development and those taken at others may evaporate for an uncomplicated reason, namely, that the measurements start to clump together. This clumping might arise because the measurements reach some ceiling (or floor) on a scale of measurement (see Fig. 2). For instance, a correlation between the performances of children on a simple test of addition at one age and their performances on the same test at later ages might drop to zero because a point is reached when all the children get full marks.

B. Changes in Relationship

An absence of a correlation between two measures as determined by a straightforward statistical test need not necessarily mean that there is no relationship. A change from a linear to an inverted U-shaped relationship might easily be misinterpreted as a loss of any relationship. Consider a case in which an intelligence quotient is repeatedly obtained from a group of

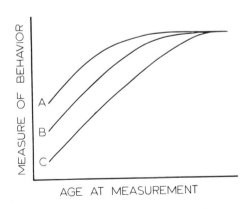

AGE AT MEASUREMENT

Fig. 2. A demonstration of a "ceiling effect." Initially individual A ranks higher than B, which ranks higher than C. The rank order on a given measure is lost at a certain stage in development because it is impossible any longer for A to show its supremacy on the scale of measurement.

children. Let us suppose that the test is one that is especially appropriate for a certain age range and contains questions like "Continue the series O,T,T,F,F . . ." It is alleged that bright children have no great difficulty in providing the answer: "S,S,E . . ." Bright adults, however, rack their brains looking for complex rules that might provide some meaningful sequence and totally miss the point that the series consists of the first letters of One, Two, Three, Four, Five . . . Now it is quite possible that children who do very well on an IQ test at one stage subsequently switch to a more adultlike problem-solving strategy before their peers and consequently do relatively badly on what is for them an inappropriate test. So an inverted U-shaped relationship between IQ at the two stages might easily arise. The change in the children's strategy would, of course, be interesting but would not signify that a fundamental *reordering* of the children had occurred as a result of some event in development. Clearly, one can look for this possibility by plotting the data in graphical form. Furthermore, with appropriate measures of behavior at later stages in development, it should be possible to show that the original rank order was eventually regained.

C. Separate Determination of Timing

If we imagine a cohort of individuals growing older together, their rank order on a particular measure at Stage 1 in Fig. 3 might correspond closely to their rank order at Stage 3. For instance, the factors responsible for the difference in their height at Stage 1 might be the same set of factors responsible for the difference at their height at Stage 3. However, a quite different set of factors might be responsible for the timing of a growth spurt. Consequently, the rank ordering of height at Stage 2 during the period of the growth spurt might be quite different from that of Stage 1 and Stage 3. Such an effect should be detectable from the data, provided, of course, that the cohort has been sampled at appropriate points during development.

D. Self-Stabilization

Mechanisms that might bring a developing system back onto a pre-determined track after it has been perturbed from that path could easily generate apparent discontinuities in development (e.g. Bateson, 1976a). Consider the case in Fig. 4, where the top-ranking Individual A suffers some trauma and for a while ranks lowest. So long as the effects of the trauma last, measurements taken before it will not correlate with measure-

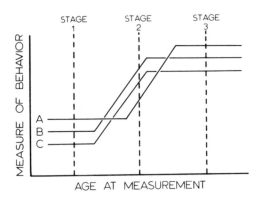

Fig. 3. At Stages 1 and 3, Individual A ranks higher than B, which ranks higher than C. However, at Stage 2, Individual A ranks lowest because the developmental change in the value of the measure is determined independently of the original rank order on that scale of measurement.

ments taken afterward. Once the individual has overcome the effects of the trauma, measurements taken after the recovery would not correlate with those taken while it ranked lowest. Mechanisms that might produce such effects are not difficult to model in principle (see Bateson, 1976a), but empirically they are difficult to identify. Nevertheless, the models raise a series of analytical questions, such as what kinds of factor can produce disturbances in the rank order and what kinds of conditions must be fulfilled for the original rank-ordering to be restored. Once those conditions are identified, it becomes possible in principle to cut the feedback required by a self-stabilizing system. This approach might be especially applicable in the case of social behavior in which the error-correcting feedback operates through a parent or a social companion.

E. Lack of Equivalence in Behavior

One reason for employing the correlational approach to discontinuities is that it sidesteps the problem of whether or not new behavior patterns really are new. But it creates a fresh problem, which is whether the "old" behavior patterns really are the same old ones when measured at later stages of development. The investigator may have been deceived by superficial descriptive similarities between different behavior patterns. A mistake of this type might be especially likely when the measure in a standard test was something like "Latency to approach" or "Time near object." Careful observation of the descriptive structure of the motor patterns might indicate that quite different systems of behavior were involved at the different stages of development. It is worth noting, though, that the investigator's deception might still be interesting since it could reveal something about changes in the organization of behavior occurring at a particular stage of development.

For instance, when domestic chicks approach a novel visually conspicuous object, they may be responding socially at 1 day old but asocially exploring it when they are 1 week old. One can investigate the character of the discontinuity by examining not only the detailed form of the behavior but also the consequences for the animal of its response before and after the change.

F. Change in Control

An issue related to the last is change in control of an unchanged motor pattern that might occur at a certain stage in development. Rosenblatt (1976) has described how 3-week-old kittens with their eyes recently open approach the mother at a distance for suckling but how when they reach her, they search for a nipple with their eyes closed. Gradually, vision comes to trigger not only the approach to the mother but also the nipple-searching behavior patterns that were formerly triggered by nonvisual sensory systems. Such a change in control might easily be associated with a change in the rank order on the scale of measurement of the behavior. In this case, the discontinuity can be investigated by an examination of the conditions that elicit the behavior.

G. New Sources of Variation

The direction of a moving billiard ball changes when it hits another ball or the edge of the table. Similarly, the developmental course of an animal may be dramatically influenced by fresh events. Unless the animal is equipped with powerful self-stabilizing mechanisms, a new source of variation can swamp the effects of previous sources of variation. This point is so

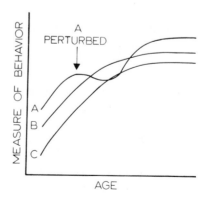

Fig. 4. At the outset, Individual A ranks highest on the scale of measurement, but after being perturbed, A temporarily sinks to the bottom of the rank order. It reaches the top again as a result of self-stabilizing mechanisms (see Bateson, 1976a). An apparent discontinuity could arise at a particular stage of development if perturbation was more likely to occur at that stage.

obvious that it needs to be emphasized. Consider a group of individuals that reach a stage in development when an aspect of their behavior is particularly prone to modification. The extent of the modifying influence need not be correlated with the rank ordering of the individuals beforehand. If the new source of variation is external, a low-ranking individual may fortuitously receive much more stimulation than a high-ranking individual. What is worse, the ways in which the old and the new sources of variation interact may be complex and not easily predictable (e.g., Denenberg, 1977). Consequently, the new rank order on a particular measure of behavior would not be correlated with the old rank order (thereby producing a discontinuity), and it might not be correlated with a measure of the events producing the change. Finally, to add to this alarmingly complex prospect, the new source of variation may be internal and not readily observable.

When processes involving elaborate interactions are suspected in development, a systematic attempt to study them must begin with a preliminary catalog of all the major sources of variation. In the case of imprinting in birds, for example, four factors are of undoubted importance: the age of the animal, the length of exposure to the imprinting object, the characteristics of the object, and the nature of experience with other objects before and after the period of imprinting (Bateson, 1977). In order to predict the ways in which such factors will interact, it is necessary to move toward explicit models of what is likely to be happening, because it is not easy to juggle mentally with all the known sources of variation and operate by intuition alone. The major benefit of working models is that the values of the interacting factors can be systematically varied. This manipulation is particularly easy if the simulation is run on a computer. Effectively, thousands of experiments can be run at little cost or trouble. This is important when most of the experiments generate either negative or trivial results. I tried this in the case of a model of imprinting and much of the time was not surprised (Bateson, 1977). However, by blindly exploring the effects of different parameter combinations, I found that the simulations did generate results that I had not expected. After some thought, the results did not seem so counterintuitive as I had first found them to be, and they convinced me of the value of the approach. The simulations pointed to areas of ignorance and certainly forced me to lay bare my assumptions and formulate the empirical questions much more precisely than I had done beforehand. This was the payoff, not the model itself, which I regard as being eminently disposable.

Although a new source of variation has been presented here as though it were a different explanation for a discontinuity in development from changes in the organization of the behavior, it could clearly be responsible for that reorganization. Indeed, a change in the character of behavior given in a particular context or a change in the control of a particular pattern of behavior must arise from *some* new source of variation. So this final explanation for discontinuities is essentially complementary to the other two (namely, Sections E and F).

V. CONCLUSION

The most important message that can be drawn from studies of behavioral development so far is that all sorts of genetic and environmental factors contribute in all sorts of different ways to the distinctiveness of a particular pattern of behavior. These determinants are all on the left-hand side of a gigantic equation and need bear no straightforward set of relationships to the products on the right-hand side. The systems involved in development are elaborate, and the underlying processes are only just beginning to be analyzed. However, to argue that the developmental processes are complex is *not* to argue that they can never be understood. The major sources of variation can be systematically located, the ways in which they interact can be modeled, and the models can be tested. Despite the undoubted complexity of behavioral development, general explanatory principles are, I believe, within our reach.

In order to make the reach for principles somewhat less haphazard than it has often been in the past, I have argued that we need to run competitions between the alternative interpretations for specific sets of data. To illustrate how this competition might be conducted, I have taken discontinuities in behavioral development, an issue that excites considerable interest at the moment, and have offered a variety of different explanations. The phenomena are undoubtedly heterogeneous, and so while each explanation could apply some of the time, it is extremely doubtful whether any one of them will apply all of the time. Some of the explanations point to deficiencies in method; some imply that after a loss of the original rank order on a particular measure, there will be a recovery of that order, and some suggest that a change in rank order is permanent. The analogy with metamorphosis employed by Kagan (this volume) would be relevant only to phenomena brought about by the last group of processes. Clearly, it is not difficult to devise ways of distinguishing between many of the possibilities. The merit of listing them lies, therefore, in the formulation of crisp questions—and developmental research could do with some crisp questions. Admittedly, the explanations are not always mutually exclusive. It is entirely plausible, for example, that a self-stabilizing system is imperfect, so that a perturbation in development stimulates a partial recovery and also becomes a new source of variation (see Bateson, 1976a). Nevertheless, if adequate systems theories of behavioral development are to emerge, it will be important to pit the hypotheses against each other in specified contexts.

Finally, it is worth noting that clear-headed experimenters are much more likely to invest effort in testing theoretical predictions if they feel that by doing so they are likely to settle an issue.

VI. ACKNOWLEDGMENTS

I am particularly grateful to Judy Dunn and Robert Hinde for their comments on an earlier draft of this manuscript.

VII. REFERENCES

Bateson, P. P. G. (1976a). Rules and reciprocity in behavioural development. In Bateson, P. P. G., and Hinde, R. A. (eds.), *Growing Points in Ethology*, Cambridge University Press, Cambridge, 548 pp.

Bateson P. P. G. (1976b). Specificity and the origins of behavior. *Adv. Stud. Behav.* **6**:1–20.

Bateson, P. P. G. (1977). Early experience and sexual preferences. In Hutchison, J. B. (ed.), *Biological Determinants of Sexual Behaviour*, Wiley, London.

Clarke, A. M., and Clarke, A. D. B. (1976). *Early Experience: Myth and Evidence*, Open Books, London, 314 pp.

Denenberg, G. H. (1977). Paradigms and paradoxes in the study of behavioral development. In Thoman, E. B. (ed.), *The Origins of the Infant's Social Responsiveness*, Erlbaum, Hillsdale, N.J.

Dunn, J. (1976). How far do early differences in mother–child relations affect later development. In Bateson, P. P. G., and Hinde, R. A. (eds.), *Growing Points in Ethology*, Cambridge University Press, Cambridge, 548 pp.

Eibl-Eibesfeldt, I. (1970). *Ethology: The Biology of Behavior*, Holt, Rinehart & Winston, New York, 530 pp.

Gottlieb, G. (1976). The roles of experience in the development of behavior and the nervous system. In Gottlieb, G. (ed.), *Studies on the Development of Behavior and the Nervous System, Vol. 3, Neural and Behavioral Specificity*, Academic Press, New York. 352 pp.

Isaac, D. J., and O'Connor, B. M. (1976). A discontinuity theory of psychological development. *Human Relations* **29**:41–61.

Lehrman, D. A. (1970). Semantic and conceptual issues in the nature–nurture problem. In Aronson, L. R., Tobach, E., Lehrman, D. S., and Rosenblatt, J. S. (eds.), *Development and Evolution of Behavior*, Freeman, San Francisco, 656 pp.

Lorenz, K. (1965). *Evolution and Modification of Behavior*, University of Chicago Press, Chicago, 121 pp.

Rosenblatt, J. S. (1976). Stages in the early behavioural development of altricial young of selected species of non-primate mammals. In Bateson, P. P. G., and Hinde, R. A. (eds.), *Growing Points in Ethology*, Cambridge University Press, Cambridge, 548 pp.

Schneirla, T. C. (1965). Aspects of stimulation and organisation in approach/withdrawal processes underlying vertebrate behavioral development. *Adv. Stud. Behav.* **1**:1–74.

Scott, J. P., Stewart, J. M., and De Ghett, V. J. (1974). Critical periods in the organisation of systems. *Devel. Psychobiol.* **7**:489–513.

Chapter 4

CONTINUITY AND STAGE IN HUMAN DEVELOPMENT

Jerome Kagan

Harvard University
Cambridge, Massachusetts 02138

Although change is as common as continuity in morphological or psychological development, the psychologist is particularly friendly to the latter idea and assumes, unless shown otherwise, that the psychological structures formed by early experience remain untransformed. This mental set is also applied to rate of development, for many developmental psychologists have tried to show that infants who attain the maturational milestones of object permanence, walking, or the first spoken word earlier than others retain that precocity for several years or longer, despite the fact that entomologists have not suggested that different rates of morphogenesis from larva to pupa to butterfly are of much consequence in explaining variations in the functioning of the adult form. Since there is little robust evidence favoring either the maintenance of early psychological structures or initial precocity, it is reasonable to ask why the Western psychologist has been so receptive to arguments for continuity.

I. THE MEANINGS OF CONTINUITY

The concept of continuity bears a complex relation to the notion of identity and constancy and to assumptions about the ratio of frozen to dynamic moments in nature's objects. Is the 6-inch yellow pencil lying motionless on my desk at 10:05 the same pencil that was there at 10:00? That may seem like a silly question, but, as Bertrand Russell has noted, it is just such questions that often veil profound enigmas. If I believe the pencil to be an envelope of quarks interacting with each other and the surround, then clearly its properties at 10:05 cannot be the same as those that existed a few minutes earlier. That conclusion usually provokes the irritated but confident rebuttal that despite the invisible alterations, the pencil has not changed across the 5-minute interval. It is still yellow, still capable of making marks on paper, and still 6 inches long. It has retained its identity because these qualities have displayed continuity. Since both statements are true, it is clear that the meaning of identity and, by implication, continuity depends on the perspective assumed. When we move from pencils to living systems, the need for a relativistic stance becomes a little more obvious. And when we add psychological qualities to the domain of consideration, the importance of perspective becomes overwhelming, for statements about constancy and change depend on the level of analysis and, therefore, the language of description.

Biologists and psychologists acknowledge that although the organism never stops changing, rate of change varies considerably across qualities. When the rate slows to some critical value, the scientist is prepared to invoke the word *constant*. Body weight displays a greater rate of change during the first 15 years of life than it does after that time; it is at the point of slowing that we say a person's weight has attained some degree of constancy. When the amount of variation in a quality is small relative to its average value, we are inclined to attribute constancy to that quality. But the size of the ratio we treat as diagnostic of constancy varies with the characteristic. We demand that the ratio be very small for stature or for response time in the pressing of a button to an auditory signal. We accept a much larger ratio for IQ scores, for a variation of 8 IQ points over a one-year interval for a child with an IQ of 100 is regarded as unimportant. Each domain has unstated understandings regarding the ratio of variation to average value that is indicative of constancy.

But this ipsative method of evaluating the stability of a quality is not often used for psychological dimensions, where relative rank on test scores or frequency of behavior is the datum. The common procedure is to evaluate the relative positions of a sample of individuals on some variable over time and regard the resulting correlation as the index of constancy.

Let us compare these two strategies for making judgments of constancy for a particular dimension, say, the number of unrelated words a child can recall. A verbal-recall task is administered to a Boston middle-class child every month from 3 through 15 years of age in order to determine at what age, if any, the child's recall score attains constancy. That determination depends, in part, on how the data are organized. If the data are examined month by month, we note a slowly rising curve; but if the data are grouped by two-year intervals, we note a sharply rising curve that plateaus at about 8–9 years of age. If the ratio of change to average recall score is 1 to 7 at the plateau, we are likely to assume constancy of that competence for that child. But suppose we perform the same study with 19 additional children and find that among these 19 there is no year-to-year variation in their performance after age 7. All 19 always remember seven words. We might now regard the performance of our original child as less constant and retract the original assignment of stability to his memory performance. A person can come to display less change in some quality, relative to his past behavior, and yet be more variable than members of a cohort with whom he is being compared.

Child psychologists often quantify the number of seconds of vocalization an infant displays to a set of visual stimuli. If a set of absolute vocalization times varied dramatically from occasion to occasion, but the children retained their same rank, psychologists would conclude that vocalization was a stable quality, despite the fact that ipsative analyses revealed high ratios of variation to the mean over time. Chinese infants are consistently less vocal to visual and auditory events than are Caucasians over the period 7 through 29 months of age, even though vocalization increases in a major way for both ethnic groups over this same interval. The stability correlations for vocalization among Chinese children are moderately high during this period, despite the sizable intraindividual variability over time (Kagan, Kearsley, Zelazo, and Minton, 1976). A more common practice in studies of long-term stability of human traits is to correlate two sets of ranks on two phenotypically different qualities. In this case, an ipsative analysis is impossible. For example, a group of infants are ranked on mean length of utterance at 18 months and on verbal IQ at 5 years. Although the metrics for length of utterance and IQ are different, if the children retained their relative ranks, psychologists would be tempted to conclude that some hypothetical quality remained stable. Many might call that quality "intelligence."

Developmental theorists are fond of making predictions about the future form of some early behavior that they assume undergoes a lawful transformation. Although the theoretical prediction implies an ipsative analysis, the empirical implementation is always normative. For example, one popular hypothesis states that a child who displays signs of fear following maternal separation at 1 year will be dependent on his mother at age 5. In order to verify this prediction, one must follow a group of children longitudinally and demonstrate that those who displayed more distress than their peers at 1 year displayed more dependency than their peers at age 5. If this result emerged, the investigator would conclude that a quality akin to "insecurity" was stable. But that statement requires a referent group, since almost all infants show distress following some maternal departures, and all 5-year-olds display some dependency. Moreover, a change in the comparison group at age 5 might alter the decision about stability. A child might be high in the rank order for both separation distress at 1 year and dependency at age 5, relative to one referent group, but if the referent group at age 5 was altered, he might have a markedly different position. Consider a real rather than a hypothetical example. At both 20 and 29 months of age Chinese children are more inhibited and wary with an unfamiliar peer than Caucasians. Hence, the correlation between signs of fearfulness at 20 and 29 months for the pooled group of Caucasians and Chinese is moderately high, and one is tempted to conclude that the trait of apprehension is stable over the 9-month period. But if the correlations are run separately for Caucasians and Chinese, there is no stability of inhibition within either group. If a child in a Boston school had reading scores at the 30th percentile over a two-year period, his reading ability would be regarded as stable. But if the child were moved to a rural Kenyan school he would probably be reading at the 70th percentile, despite no real change in his ability.

The meaning of continuity of a psychological quality based on normative correlational analyses is to be differentiated both from the logical meaning of identity and the constancy of a quality based on an ipsative analysis. The confusion in the uses of these terms has made the study of psychological continuity ambiguous and vulnerable to justified criticism. When statements about continuity are based on sets of ranks, we must always add a statement about the qualities of the referent group with whom the child is compared. It is necessary, therefore, to distinguish among (1) variation in the rate of change of a particular quality within a single individual; (2) relative variation in the rate of change when a particular individual is being compared with a referent group; (3) maintenance of a set of ranks on a variable whose metric remains constant, and (4) maintenance of a set of ranks for variables with different metrics. Most psychological research on continuity falls into one of the last two categories.

II. CONTINUITY IN STAGE THEORY

The desire to affirm developmental continuities flies in the face of obvious changes in the child's behavior and talents over the first dozen years, changes so dramatic that the concept of *stage* was invented. Since the concept of stage implies discontinuity in both deep and surface structures, many stage theorists have satisfied the continuity axiom by invoking still another meaning to that term, namely, the meaning implied by epigenesis. This notion states that the competences gained during one period of development are incorporated into those of the succeeding period. This hypothesis permits one to retain a faith in the continuity of early structures while accommodating to the discontinuities in behavioral performances.

Let us examine this idea in the context of the infant's reaction to and processing of stimulus information. The ability to detect an event as discrepant from one's schema emerges very early and is a major influence on the 3-month-old's selective distribution of attention to objects and people in his life space. But six months later, a new competence appears. The 9- to 10-month-old retrieves representations of past experiences and generates predictions in the face of discrepant events. As a result, he shows object permanence and distress following separation from his caretaker. Obviously, if the child cannot detect a discrepancy, he will not display the new competence. But there is no good reason to assume that the ability to detect discrepancy is part of the competence involved in the ability to retrieve a schema, compare it with the present, and generate relations between them. The child must be able to support and coordinate his head and arms if he is to walk without falling down. In that sense, the earlier mastery is required for the later one. But the mechanisms that permit control of head, arms, and trunk are not the central ones involved in walking. The simple zygote eventually develops a notochord. Although that structure was a potential in the genome, it is not derivative of any structure in the zygote, even though the zygote had to preexist.

The abilities to speak a language of increasing complexity, to experience shame or guilt, to operate on several bits of information, and to examine the logic of a set of propositions are inserted into the ongoing processes of growth like new routines in a complex computer program. Each routine is inserted at the proper time, and a temporal order is preserved. The routines may consist of unique abilities that are not derived from former competences but, in conjunction with the former, yield a new level of talent.

When Piaget suggests that concrete operations are derivative of sensorimotor schemes, he implies that the concrete operations are transformed derivatives of sensorimotor structures. But it is not obvious how the operation called *multiplication of classes* is a derivative of any sensorimotor competence. The earlier competence may necessarily appear before the latter, but the latter is not, in a strict sense, an outgrowth of the earlier one. Growth is not a relay race with batons being handed from one runner to another but more like a painting. In no sense does the initial row of trees on a canvas give rise to the outline of a figure reclining by the trees, even though the latter may have followed the former.

McCall *et al.* (1976) has performed an interesting analysis on a longitudinal set of infant test data administered to the Berkeley longitudinal sample. The data consisted of early forms of the Bayley scale administered monthly from 1 to 15 months every three months until 30 months and semiannually until 5 years. McCall performed a principle-component analysis on the data and correlated the principle-component scores across age.

The data revealed major discontinuities at 3, 8, 13, and 20 months. The intraindividual cross-age correlations for scores on the items that loaded on the principle components dropped at these points. The correlations between adjacent ages within developmental eras (e.g., between 3 and 6 or even 12 months) were higher.

A similar result occurred in a longitudinal analysis of attentiveness, vocalization, and smiling to visual and auditory events in a group of children seen at 4, 8, 13, and 27 months (Kagan, 1971). There was very little intraindividual stability from 4 months to 8, 13, or 27 months or from 8 to 27 months, but there was moderate stability between 8 and 13 months and between 13 and 27 months, in accord with McCall's findings.

These two corpora of data, with different populations and dependent variables, imply that major psychological changes occur around 8 months of age and again after the first birthday.

There are invariances of order due to the operation of genetically controlled mechanisms, but there need not be closely dependent relations between all the structures of successive stages. It may even be that when new competences are added to the pattern of existing abilities, earlier ones vanish. Once the young child realizes the class to which the dog next door belongs, he may never again be able to react to any dog as if it were outside of any category. When blastula becomes gastrula, the bundles of competences of the former stage are terminated. It was used, like scaffolding for a building, and is not needed any more.

It may be useful, therefore, to question the strict continuity of competences and the strong form of the epigenetic view. As there are discontinuities in phylogeny—man's language ability is not an obvious derivative of the baboon's signaling behavior—or in the growth of a plant from seed to bud to blossom, so too there may be more discontinuities in psychological development than we have wanted to acknowledge. The state of being a caterpiller always occurs before butterflyhood, but it is difficult to discern the competences that the adult inherited from the hairy worm. In declaring a faith in strong psychological continuity from stage to stage, we may have been committing the ancient error of *post hoc ergo propter hoc.*

III. THE HISTORICAL BASES FOR FAITH IN CONTINUITY

The assumption of a thick cord of connection between stages, a contingent relation between earlier and later structures, is attractive to the

Western mind because we like the notion of unitary deep structures that explain phenotypically diverse performances. It is less pleasing to consider successive classes of behavior, each resting on its own bottom. Modern stage theories satisfy our deep longing for a hidden unity that weaves experience into a seamless fabric.

The presence of an unbroken relation between infancy and later childhood can be traced to Plato's assumption in the *Timaeus* that all events have prior causes and to the arguments proposed by medieval philosophers to affirm the existence of God. Thomas Aquinas's defense of the idea of God bears a strong resemblance to contemporary arguments for the origins of adult personality in infant experience. Aquinas accepted the Greek premise that all dynamic events had a prior cause. Hence, if one traced all movements back to the original incentive, one would arrive at the original unmoved mover, which, for Aquinas, was God. Psychologists who believe that adult behavior is a partial derivative of early experience tacitly suppose that some adult dispositions must have a connection to earlier ones, the earlier ones to still earlier events. In this regress, one arrives at the nursery convinced that in a newborn's thrashing or lethargy one sees the origins of antisocial behavior (Yang and Halverson, 1976).

There is phenomenological support for faith in connectedness. We reflect on our past and have the illusion of a unity. The illusion is based partly on our need to seek coherence in our own lives and partly on a palpable pressure to see rationality and causality in decisions.

Additionally, we do not like to see anything wasted or thrown away; perhaps that is why the principle of conservation of energy is so aesthetic. The possibility that the intensity, variety, and excitement of the first years of life, together with the extreme parenting effort that those years require, could be discarded like wrapping paper that is no longer needed is a little threatening to our Puritan spirit. We want to believe that it is possible to prepare the child for the future as one saves for a rainy day. Application of that maxim to psychological development leads to the deduction that if children are treated optimally during the early years of life, the healthy beliefs and behaviors established during that first period will be adequate protection against later traumas. This idea may be true, but there is no commanding proof of its correctness.

Psychoanalytic theory received a warm reception when it was introduced into the United States during the third and fourth decades of this century. Until that time, the infant was seen as a socially isolated organism impelled by inevitable maturational forces. Psychoanalytic theory placed the child in a social matrix in which the parents' actions became a central determinant of the child's development and a phenomenon to study in its own right. It is possible that one reason psychoanalytic theory was more popular in the United States than in continental Europe was that America wanted to believe in the infant's susceptibility to parental influence. American families were ready to celebrate a theory of development that made early family experience primary and treated infancy as the optimal time to mold the child.

A. Evolutionary Theory and Infant Determinism

Although the belief that the structures laid down by experience during the first three years of life had a primacy not easily muted was strong prior to the 19th century, evolutionary theory strengthened this idea considerably.

Darwin reclassified the human infant as a member of the category *animal* by positing a continuum in evolution. Since animal behavior was regarded as instinctive, inflexible, and therefore resistant to change, a paradox was created. How was it possible for man to be varied in custom and habit—so flexible and progressive—if he were such a close relative of creatures whose behavior appeared to be excessively stereotyped and rigid? One way to resolve the dilemma was to award a special function to what seemed to be the more prolonged period of infant helplessness in humans as compared with animals. Since most assumed that all qualities of living things had a purpose, it was reasonable to ask about the purpose of man's prolonged infancy. John Fiske, among others, argued that infancy was a period of maximal plasticity, the time when adults were to teach children the skills and ideas that they would carry with them throughout life.

In a lecture at Harvard in 1871, Fiske noted that man's power to control his environment and to enhance progress had to be due to his educability—a potential that Fiske believed animals lacked. How, then, to explain why man was so educable? Following popular rules of inference, Fiske looked for other major differences between man and animal, and of the many candidates he could have selected—language, the opposable thumb, upright stature, an omnivorous diet, sexual behavior throughout the year—he selected the prolongation of infancy. The logic of the argument consisted of two premises and a conclusion. Nature had to have an intended purpose for the initial three years of human incompetence and dependence. Since the most likely purpose was to educate the child, it must be the case that the child is maximally malleable to training during that early period. That conclusion was congruent with a deep belief in continuity of character from infancy to later childhood, it served to keep parents self-conscious about their actions with their babies, and it was an argument for building good schools. It also provided an experiential explanation for individual differences in adult success—which was attractive to a democratic and egalitarian society (Fiske, 1883).

America and England, the two major Protestant nations in the Western community, wanted to believe in the eventual attainment of an egalitarian society where, if conditions of early life were optimal, all citizens potentially could attain dignity and participate effectively in the society. Indeed, during the years just prior to the American Revolution, clergy and statesmen wrote impassioned essays stressing the importance of early family treatment and proper education in the prevention of crime and the safeguarding of democracy. Despotism could be eliminated if all children were well nurtured and properly educated—a conviction not unlike Plato's assumption that if one knew what was good, immorality was impossible. From America's birth until the present, a majority have believed that the correct pattern of experiences at home and school would guarantee a harmonious society. This view, which was in accord with the doctrine of infant malleability, invited each generation of parents to project onto the infant their political hopes for the future. Every infant was a fresh canvas, and the community was ready to receive a psychological theory that would make social experience the major steward of growth.

IV. EVIDENCE FOR CONTINUITY IN HUMAN DEVELOPMENT

Recently published data have led some psychologists to question the traditional belief in the long-term effects of early experience on future behavior. For the first time in the modern history of child psychology, one senses the beginning of some skepticism toward the strong form of the continuity assumption. The belief in psychological continuity within an individual life led, during the 1930s, to the initiation of several long-term longitudinal studies whose goal was to document the continuity that many were sure had to be present. The twin assumptions that the infant is malleable to experience and that early structures persist for decades, which were seen as necessarily correlated, are now regarded as independent. If the young child is malleable in his current environment, then a change in context might alter the profiles produced by the prior context. The 19th-century commitment to the malleability of the infant is being taken seriously, for the young child seems to have an extraordinary capacity for change, given new adaptation demands. That insight was disguised in the past, perhaps because we implicitly assumed that the average child's environment would not—or could not—change. In a more stable America, most children remained in the same family context and neighborhood until it was time to embark on adult responsibilities. The continuity of behavior noted by relatives and friends was attributed to forces within the child rather than to the environmental context in which he played out his roles. If a small marble placed in a trough on an incline rolled unerringly in a straight line, one would not credit the marble with the capacity to maintain continuous linear movement. We did not apply that principle to development but concluded instead that the child had a mysterious power to sustain his psychological direction without the help of a guiding track.

The results of the first longitudinal studies, published during the last 15 years, provide surprisingly fragile evidence for strong forms of continuity from infancy forward. The Fels study, which involved Caucasian families from southwest Ohio, did not support the notion that infant traits like fearfulness, anger, motoricity, or dependency predicted theoretically relevant dispositions in adolescence or early adulthood. It was not until preadolescence that consistent predictive validity from childhood to adulthood appeared (Kagan and Moss, 1962).

It is still not clear whether there is any strong continuity from the period of infancy to later childhood, but the burden of proof has shifted from those who deny continuity to those who claim that it is present. What does seem clear is that the child does not carry a set of "qualities" through development that is independent of the context in which he lives. If there is continuity of dispositions from infancy to adolescence, it is likely to appear under conditions where the environment promotes or maintains the infant's particular qualities. For example, some 4-month-old babies babble a lot and some are relatively quiet, and this seems to be a biologically based disposition. If a babbling baby is born to a middle-class well-educated parent, two years later that child is more talkative than a quiet 4-month-old born to parents of the same social class. But a highly vocal 4-month-old born to lower- or working-class parents, who generally do not encourage early vocalization, was not any more talkative than a low vocal infant (Kagan, 1971). The infant's qualities are continually adapting to environmental pressures.

A principle that seems to emerge from the longitudinal studies is best termed *growth toward health*. The child tries to grow toward adaptation, if the environment will permit him. The mistaken assumption made by the early investigators was that a node of pathology in infancy was like a blemish and, like Lady Macbeth's bloody hands, could not be wiped clean. Dispositions that are maladaptive or pathological can be outgrown, while adaptive characteristics are more likely to be retained. Macfarlane (1963, 1964) has reported that many children and adolescents who showed pathology lost it as they found better environmental niches and opportunities that permitted old anxieties to be resolved. This generalization is also supported by the early studies of Wayne Dennis (1938), who restricted a pair of twin 2-year-old infants for most of the first year. After they returned to a normal environment, they eventually began to display a growth function that was normative.

A. Some Recent Data

A recently completed longitudinal study of Chinese and Caucasian children studied from 3 to 29 months of age provides additional data relevant to these themes. Fearfulness has been one quality that some presumed to be stable over a period of at least several years, with family experiences contributing to the stability of that disposition. The data we have gathered not only fail to support that idea but also indicate that competences associated with stages in cognitive development may explain the times of emergence of these fears.

We are concerned with two sources of apprehension that appear during the first two years of life. The distress displayed when a child watches a parent depart (called *separation fear*) and apprehension on encounters with an unfamiliar child. We assessed the first construct by having the mother leave the child alone or with a stranger and observing the incidence of crying and inhibition of play. We assessed the second introducing an unfamiliar peer into a situation where the child had been playing happily and by coding inhibition of both play and vocalization. A decrease in either or both responses was interpreted as a sign of wariness.

The children were 95 Caucasian and Chinese children, some of whom were attending a day-care center; others were being reared at home. These children were observed in a standard separation situation at $3\frac{1}{2}$, $5\frac{1}{2}$, $7\frac{1}{2}$, $9\frac{1}{2}$, $11\frac{1}{2}$, $13\frac{1}{2}$, 20, and 29 months and with an unfamiliar peer at 13, 20, and 29 months. The data reveal, first, that there was a delimited and relatively fixed period when signs of apprehension at these incentive events appeared. After this interval, the wariness vanished. Separation anxiety peaked from 9 to 15 months of age and peer apprehension at 13 and 20 months. Further, there was no long-term stability of the disposition to cry in the separation situation (see Table I) or to inhibit play or vocalization in the encounters with the unfamiliar peer.

Additionally, there was no relation between the frequency of crying to separation and signs of apprehension related to the unfamiliar child (Kagan *et al.*, 1976).

B. A Suggested Interpretation

We believe that a key to understanding the emergence of these fears toward the end of the first year is the recognition of a new ability to retrieve a schema from memory and to compare the retrieved structure with a present event. The child of 8 months, unlike the younger infant, is able to retrieve, not just recognize, schemata for events that occurred more than a few seconds earlier (Kagan, 1976). The child can generate a structure for a

Table I. Cross-Age Stability of Distress to Separation
(Phi Coefficients)[a]

Age	3	5	7	9	11	13	20	29
3	—	16	21	20	05	04	30	16
5	20	—	00	21	06	14	03	22
7	36	00	—	32[b]	07	18	24	20
9	06	02	17	—	28	03	05	32
11	03	17	04	31	—	28	37[b]	16
13	06	08	04	07	16	—	56[c]	16
20	00	01	06	16	47	15	—	30
29	37	33	29	14	39	49[b]	24	—

[a] Chinese to the right and above the diagonal; Caucasians to the left and below the diagonal.
[b] $p < 0.05$
[c] $p < 0.01$.

past event that is not in the perceptual field. The suggestion that a memorial capacity is central is supported by the results of a study that varied the delay between the hiding of a toy and allowing the child to reach for the toy in the standard "A not B" paradigm. The child was allowed to obtain the toy at locus A on two trials. The child then watched the examiner hide the toy at locus B. If the delay between the hiding at B and allowing the child to reach for the toy was 7 seconds, the 7- and 8-month-olds went to position A; if the delay was 3 seconds, they did not make the error and went correctly to position B. But no 10-month-old made the error under either 3- or 7-second delays, suggesting the growth of increased recall capacity between 8 and 10 months.

Millar (1974) has invented a clever procedure that supports the hypothesis of a change in memory competence toward the end of the first year. The infants, 6 or 9 months old, sat in front of an aluminum canister manipulandum. If the child hit the canister in front of him, he was rewarded with sounds and changes in light, which occurred in front of him. For some, a very distinctive green plastic ring circumscribed the area where the reward would appear. For others, the ring was absent, and hence the source of the reward was not visible. The 9-month-olds banged the canister under both conditions, whether or not the green ring was present. The 6-month-olds increased their rate of striking the canister only if the green ring was there, not when it was absent. These data imply that the 9-month-olds were activating structures representing the reward that might be delivered if the proper behavior was issued, while the 6-month-olds were too immature to activate or remember that idea and required a reminder in the form of the green ring in order to be motivated to hit the canister (Millar, 1974). Millar wrote, "Six month olds, in contrast to 9 month old infants, are incapable of making the spontaneous use of centrally held information and relating it to ongoing activities" (p. 515). Schaffer (1974) has also suggested that the new cognitive competence that emerges at this time is the ability to activate from memory schemata for absent objects and to use them to evaluate a situation:

> Initially, each stimulus is treated in isolation; and although the memory store may be checked for representations of that same stimulus, it is not compared with different stimuli or their representations. In time, however, the infant becomes capable of relating stimuli to one another. . . . As a result the strange stimulus can be considered simultaneously with the familiar standard, even though the latter is centrally stored and must therefore be retrieved. (Schaffer, 1974, p. 22)

We believe the "short-term memory" stage collapses after most experiences, especially if they are poorly articulated, in infants under 7 months of age. The structures created in the past cannot be retrieved in the absence of relevant incentive events, even though the young infant can, of course, recognize an event that is similar to a schema derived from past experience. But by 9–10 months, changes in the central nervous system have allowed the stage to remain intact for a longer period; hence, the infant can retrieve the information he transduced earlier.

The new capacity to retrieve structures for events not in the field can explain some of the fears of the first year. Following departure of the mother, the child generates from memory the schema of her presence and compares it with the present. His inability to resolve the inconsistency between those structures leads to uncertainty and fear. But why does this process elicit fear in the case of maternal departure and not in a situation when the child fails to find a toy under a cloth after watching the adult place one there?

One possibility is that the new ability to retrieve a schema of a past event is correlated with the ability to generate anticipations of the future, representations of possible events, not just schemata of past events. Anticipating the mother's absence as she gets up to leave, a future possibility, is not synonymous with retrieving a past event and comparing it with the present. Since some 1-year-olds often cry before the mother even gets to the door, they seem to be anticipating her departure.

A second possibility is that the 1-year-old anticipates an unpleasant event, pain or danger, because he has experienced such an event in the past. He cries because of a conditioned anxiety reaction. Although this has been a popular explanation for many years and one that seems intuitively reasonable, we must ask why it is that infants all over the world, even those who are with their mothers for most of the day, suddenly expect an unpleasant event to occur when they are between 9 and 13 months of age.

The distress occurs independent of the frequency with which they have experienced distress in the past when their mother was gone. We are tempted to reject this possibility because children whose mothers leave them often—for example, children in a day-care center—do not show separation distress earlier or with more intensity than those who have their mothers with them continually.

Even though we do not favor an explanation based on anticipation of past unpleasantness when the mother was absent, we believe that the 1-year-old has a new capacity to attempt to predict what might happen following an unusual event. A dynamic, rather than a static, event is more likely to elicit apprehension, for the former (a jack-in-the-box or the mother's going to the door) is more apt to provoke questions like, "What am I to do?" or "What will happen to me?" than a static discrepancy, like finding an unexpected object under a cloth. Inability to answer those implicit questions produces uncertainty. Following maternal departure the child tries to generate a cognitive structure to explain the mother's absence and cannot do so. An additional factor is relevant. It is likely that if the child does not have a behavior to issue when he experiences the uncertainty, he may be especially vulnerable to distress.

In sum, the fear following separation is a product of two related abilities: (1) the ability to retrieve past schemata and to compare them with the present in the service of resolving the inconsistency and (2) the ability to attempt a prediction of the future. Failure to resolve the inconsistency or to predict the future will produce uncertainty, and if the child has no response to make, distress mounts. Distress is more likely to appear in relation to separation than in relation to many other surprises in the child's daily life because the schema for the mother and her presence is highly salient. Since the child will also cry when presented with a jack-in-the-box or an electrically wired box that moves unpredictably, the discrepant event does not have to be an unfamiliar person or maternal departure. But the event must engage a salient or well-articulated schema. Fear does not occur with most discrepant events in the laboratory or in the child's daily encounters because the events are unfamiliar. The minute or two of familiarization in the laboratory is not sufficient to create a firm schema. Hence, when a transformation occurs, the schema for the old event is fragile, like the vanishing message produced by a skywriting plane; the inconsistency is muted, and less uncertainty is generated. If the event is salient and the schema well articulated, the difference between present and the past remains clear.

We are left with one final puzzle. Why does the presence of a familiar person or setting reduce dramatically the occurrence of uncertainty and fear upon maternal departure? And why does the presence of the mother, especially if she is close-by, reduce the likelihood of fear in reaction to many discrepant events? There are several possibilities.

The first is that there is a continuum of uncertainty and that within the continuum each child has a threshold level of uncertainty that, when crossed, leads to inhibition and crying. The presence of a familiar setting or person acts to buffer the uncertainty and to keep it below the critical threshold. The child detects the mother's absence in the home and tries to understand it, but he is less apt to become uncertain enough to display inhibition of play or crying.

A second possibility is that the context is an essential part of the child's conceptualization of any figural event and that the infant implicitly classifies all contexts into familiar and unfamiliar ones. A dynamic figural event that is discrepant in an unfamiliar context may not be so in a familiar one. This explanation is unlikely since the child sometimes reacts with inhibition and fear to a discrepancy while he is sitting on his mother's lap.

A third possibility is that the presence of a familiar person or setting provides the child with opportunities for responses to make when uncertainty is generated. Recognition of the opportunity to behave buffers uncertainty. When the mother leaves the child with the father, distress does not occur because the father's presence provides the child with a potential target for a set of behaviors that he can issue. That knowledge appears to keep uncertainty under control. This last interpretation is profoundly cognitive. We are suggesting that the infant does not have to move toward the father; he only has to know that the parent is present. The blind 1-year-old does not have to see the mother to be protected against fear; the child must only know that she is in the room (Fraiberg, 1975). Knowledge of the availability of a familiar caretaker is sufficient to hold anxiety at bay, a principle that is also true of adults.

In sum, explanation of the fear of separation at 1 year requires at least four assumptions:

1. The new ability to retrieve from memory the schemata of past salient events and to compare them with present experience.
2. Attempts to resolve inconsistencies between schemata of the past and the present event and to predict possible future events.
3. The inability either to resolve the inconsistency or to predict future possibilities.
4. No opportunity to issue a copying response to the state of uncertainty.

The fear following separation recedes at about 2 years, when the child is able to resolve the event. The child knows where the mother is or knows that she will return. His experiences have allowed him to generate a structure that resolves, mediates, or interprets the inconsistency.

C. Summary

The use of empirical data to refute or challenge presuppositions is nicely illustrated by developmental research during the last 20 years. The faith in a strong form of continuity, which is so deep in our culture, must now be defended rather than accepted as obviously true. Further, the conception of stage as consisting of the acquisition of new structures must accommodate to a broader view that includes recognition of new competences that are regularly inserted into the developmental process. If biological development continues to be a good model for psychological growth, it is likely that we shall continue to discover discontinuities in the life history of an individual. We should expect to see, at least in early development, metamorphoses that leave behind much of the psychological baggage that was carried during the early part of the journey.

Recent advances in embryology are consonant with these conclusions. Oppenheimer (1967) has argued that the three assumptions of preformation, unity of type, and recapitulation held by early 19th-century biologists obstructed progress in embryology. The latter two have direct relevance to the issue of continuity in psychological development. Like the 19th-century embryologist, who assumed the existence of an abstract, prototypic species with little variation, many child psychologists have assumed that a competence, habit, temperament, or stage of organization also exists in the abstract, independent of context. Infants and young children were described as attentive, irritable, or in the sensorimotor stage, subduing the obvious variation in that quality that was monitored by context and occasion. The growth rate of an Amblystoma embryo varies with slight changes in acidity and temperature. There is no construct, "growth rate for amblystoma," only growth rates under particular conditions. Similarly, there is no psychological entity such as attentiveness or memory independent of classes of task contexts.

The doctrine of recapitulation, which accepted the premise of a continuous chain of life from hydra to man, assumed that the early development of a form at a higher level of evolutionary complexity reflected the development of more primitive forms. The influence of recapitulation doctrine is seen in developmental psychology in both Freudian and Piagetian theory. Both theories assume that the structures of an early stage make a contribution to the structures of a later one. Unity of type and recapitulation have not been useful ideas in modern embryology. Since embryology is a reasonably good model for developmental psychology, it may be wise to take a lesson from our sister science and examine our commitments to these axioms.

A major conceptual advance over the last decade is the increasing acceptance of the interaction of biological and experiential variables. A traumatic event during infancy seems to make a difference primarily for children from economically poor environments, not for those in more secure homes. Disease or trauma in infancy seems to affect the children in poverty more seriously than those in more affluent circumstances. We seem to be in a period of synthesis between Gesell's thesis that maturation controls most of development and Watson's pronouncement that experience has all the power. One form of that synthetic statement is that maturational forces direct the basic growth function for many psychological systems that emerge during the first two years—reaction to discrepancy, amplification of memory, object permanence, and stranger and separation fears. But experience determines the age at which these competences appear or disappear and, during the period of their emergence, the intensity and frequency of display.

The analogy to age of menarche is compelling. Most females inherit the capacity for reproductive fertility, which appears between 10 and 18 years of age. The age at which it actually appears is a function of earlier nutrition and pattern of diseases. Onset of language provides an even better analogy. All children, save a small few, inherit a tendency to speak. But the appearance of the first word and the rate at which length of utterance grows varies with exposure to adult language. Middle-class American children speak by 1 year. Isolated children in a Guatemalan village like San Marcos on Lake Atitlán, who do not experience as much verbal interaction with adults, do not utter their first words until they are over 2 years old. The new trend in child development is providing an empirical basis for the old maxim that experience and the child's biology interact in producing individual growth patterns.

V. ACKNOWLEDGMENTS

Research described in this paper was supported in part by grants from the Carnegie Corporation of New York, the Spencer Foundation, and the Grant Foundation. The author was a Belding Scholar of the Foundation for Child Development during the preparation of this essay.

VI. REFERENCES

Dennis, W. (1938). Infant development under conditions of restricted practice and minimum social stimulation. *J. Genet. Psychol.* **53**:149–158.

Fiske, J. (1883). *The Meaning of Infancy,* Houghton Mifflin, Boston.

Fraiberg, S. (1975). The development of human attachments in infants blind from birth. *Merrill-Palmer Quarterly* **21**:315–334.

Kagan, J. (1971). *Change and Continuity Infancy,* Wiley, New York.

Kagan, J. (1976). Emergent themes in human development. *Am. Sci.* **64**:186–196.

Kagan, J., and Moss, H. A. (1962). *Birth to Maturity,* Wiley, New York.

Kagan, J., Kearsley, R. B., Zelazo, P. R. & Minton, C. (1976). The course of early development, unpublished manuscript.

Macfarlane, J. W. (1963). From infancy to adulthood. *Childhood Education* **39**:336–342.

Macfarlane, J. W. (1964). Perspectives on personality consistency and change from the guidance study. *Vita Humana* **7**:115–126.

McCall, R. B., Eichorn, D. H., and Hogarty, P. S. (1976). Transitions in early mental development, unpublished manuscript.

Millar, W. S. (1974). The role of visual holding cues in the simultanizing strategy in infant operant learning. *Br. J. Psychol.* **65**:505–518.

Oppenheimer, J. (1967). *Essays in the History of Embryology and Biology,* MIT Press, Cambridge, Mass.

Schaffer, H. R. (1974). Cognitive component of the infant's response to strangeness. In Lewis, M., and Rosenblum, L. (eds.), *The Origins of Fear,* Wiley, New York, pp. 11–24.

Yang, R. K., and Halverson, C. F. (1976). A study of the inversion of intensity between newborn and preschool behavior. *Child Devel.* **47**:350–359.

Chapter 5

FEEDING BEHAVIOR OF *LEMUR CATTA* IN DIFFERENT HABITATS

Norman Budnitz

Department of Zoology
Duke University
Durham, North Carolina 27706

I. INTRODUCTION

Biologists have been trying to explain observed differences in the social structure of different species by looking at the environmental constraints on those species. For example, Crook and Gartlan (1966) observed that in open habitats (e.g., the African savanna), primate species tend to be organized into groups with large home ranges, while in more closed habitats (e.g., forests), other primate species are organized into smaller groups with smaller home ranges. On the basis of these and other observations, the authors concluded that habitats place particular constraints on social structure and that animals respond in ways that biologists should be able to predict.

The original Crook and Gartlan hypothesis has been reviewed, revised, and tested by many workers (Struhsaker, 1969, 1974; Crook, 1970; Eisenberg, Muchenhirn, and Rudran, 1972; Jolly, 1972a; Clutton-Brock, 1973, 1974). However, all of these authors, because they have looked at several different species of primates, have not dealt with one very basic problem: they cannot distinguish between habitat effects and effects due to genetic differences between species. Therefore, I chose to examine the behavior of two groups of one species of prosimian primate, *Lemur catta,* that live in different habitats. By working on one species—in fact, one interbreeding population of one species—I could observe the habitat effects on the animals without the complicating interspecific or genetic effects. If clear differences in behavior could be found *within* a species, then there would be a firmer basis for making *between*-species comparisons. It is better to compare apples and oranges after the characteristics of apples and the characteristics of oranges are well known.

Several authors have reported work on one species of primate living in different habitats (Struhsaker, 1967; Gartlan and Brain, 1968; and Frisch, 1968). However, most of these comparisons were not based on quantitative data on either the behavior of the animals or the differences in the habitats. Gartlan and Brain looked at vervet monkeys (*Cercopithecus aethiops*) in two habitats, one rich and regenerating and the other poor and deteriorating. They found that the monkeys in the rich habitat were healthier and had smaller home ranges. The monkeys in the poor habitat were less healthy, had larger home ranges, and had developed two notable differences in social behavior: a signal system for long-distance communication and a more intense mother–infant bond. However, no quantitative data were presented for any of these observations.

Frisch (1968) found a high degree of variability in the behavior of individual Japanese macaques (*Macaca fuscata*) and hypothesized that this variability might have a profound effect on the behavior of the macaque troops: "While environmental influences appear to exert at best a very limited influence on the peculiar behavioral pattern of a troop, adoption of a new type of behavior by an individual has often resulted in a durable modification in the pattern of behavior characteristic of the entire group" (page 245). Again, however, no quantitative data were presented to support this hypothesis.

Struhsaker's (1967) study of vervet monkeys (*C. aethiops*) does present data on the ranging behavior of four troops living in various habitats. These data parallel the data we collected on the lemurs and are discussed below in Section IIIA on home ranges.

The lemurs of Madagascar are a very diverse group of prosimians. The adaptive radiation of these animals into approximately 20 living species may be the result of the wide variety of habitats available on the island (Jolly, 1966). I chose to look at the feeding and ranging behavior of lemurs because one of the basic environmental constraints that any animal faces is the availability and distribution of food resources and these constraints would be different in different habitats. It seems very likely that differences in social organization would be influenced by differences in feeding and ranging behavior. For example, if one troop of lemurs has to spend more time feeding than another troop, the former troop would have less time to spend on social interactions.

The reserve in which we made our observations is made up of several habitats. To look at the effects of these habitats on the behvavior of the lemurs, we first had to express the habitat differences quantitatively. *Lemur catta* are basically herbivorous. Therefore, since many of the physical constraints on the movements of the lemurs and the availability of food in the study area both depended upon vegetation, we did a vegetation analysis. The first section of this paper deals with the habitat differences revealed by this analysis.

Once we could distinguish clearly between the habitats, we could look at differences in the feeding and ranging behavior of the groups and see if these differences correlated with habitat differences. The second part of this paper deals with those behavioral correlations.

II. DESCRIPTION OF THE HABITAT

A. Location

Lemur catta is found in scattered populations in south and southwestern Madagascar. This region of Madagascar is in the rain shadow of the mountainous "backbone" that runs from north to south along the entire length of the island. East of these mountains, facing the Indian Ocean, is an extensive area of rain forest. To the west, sheltered from the prevailing easterly winds, the climate is much drier and the vegetation is characterized by deciduous forests and vast expanses of dry grasslands where these forests have been cut down for agriculture. In the south and southwest of Madagascar, the rainfall is even more reduced, resulting in a semiarid climate. The mean annual rainfall in this semiarid region varies from 350 mm to 600 mm per year (Tattersall and Sussman, 1975). The vegetation is very scrubby and is dominated by xerophytic plants of the endemic family Didiereaceae, as well as many species in the families Asclepiadaceae and Euphorbiaceae. These plants are often characterized by thorns, succulent leaves, odoriferous compounds, or latex.

Lemur catta does not live in the subdesert vegetation. Scattered throughout the southern region are gallery forests—relatively lush forests with tall, spreading trees. The gallery forests are found along rivers running through savannas and semiarid regions (Richards, 1957). They derive their moisture from the water in the ground rather than from direct precipitation. Our field study was carried out in a private reserve owned by the de Heaulme family in the town of Berenty. The reserve is just such a gallery forest along the Mandrare River. The average rainfall in Berenty is about 500 mm per year, most of this coming in January, February, and March. During the year of this study, most of the rains did not come until April. In some years, there is no rain at all.

B. Vegetation Types

Figure 1 is a map of the Berenty reserve showing the various vegetation types. (This figure also shows the boundaries of the home ranges of the two troops that we studied most extensively.) Note that the reserve has been divided into four vegetation types. Along the Mandrare River is a continuous canopy forest dominated by *Tamarindus indica* and other large trees, such as *Celtis gomphophylla*. These trees reach heights of 20–25 m. As the distance from the river increases and as the water content of the soil decreases, this continuous-canopy forest thins out. In the open-canopy sections, the dominant trees are often tamarinds, but they are generally much smaller than their closed-canopy counterparts, averaging about 16 m in

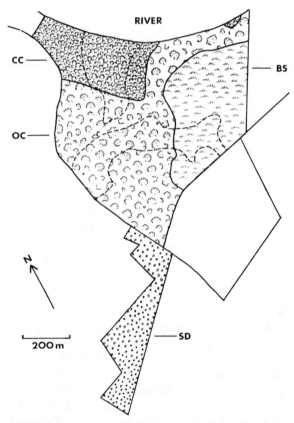

Fig. 1.   Vegetation types for the Berenty reserve, including boundaries (dashed lines) of the home ranges of Troop ST and Troop DUD (see Fig. 5). CC—closed-canopy forest; OC—open-canopy forest; BS—brush-and-scrub forest; SD—subdesert forest.

height. In many parts of this open-canopy forest some of the tamarinds are replaced by other large, spreading trees, *Nestina isoneura* or *Acacia rovumae*. Since the canopy is not continuous, there are large open areas where thorny vines, such as *Capparis sepiaria,* spread out into almost impenetrable tangles.

The third vegetation type that characterizes a large part of the reserve is what Sussman (1972) called brush-and-scrub forest. This part of the reserve is much drier than the preceding two areas and the trees are much smaller. Tamarinds are still found scattered throughout the brush and scrub, but the other plants are generally quite different. The dominant plants are bushes and small trees (*Azima tetracantha, Maerua filiformis, Salvadora angustifolia,* and *Guisivianthe papionaea*), and there is no canopy at all. A succulent vine that grows in this brush-and-scrub region, *Xerosicyos perrieri,* is an indicator species for this habitat and is quite important as a source of water for *L. catta.*

The fourth section of the reserve, the strip of land that extends to the southwest, is a small piece of the subdesert forest characterized by the Didiereaceae, Asclepiadaceae, and Euphorbiaceae described above. This type of vegetation was probably the natural boundary of the gallery forest. Now, however, most of this subdesert vegetation has been cut down and has been replaced by a plantation of sisal (*Agave rigida*).

Lemur catta lives only in the first three regions of the reserve: the closed- and open-canopy forests and the brush and scrub. We never observed them in the subdesert vegetation. Along the river to the east and west of the reserve are areas of degraded gallery forest. Since cattle were allowed to feed in these areas, the undergrowth was often devastated. *Lemur catta* did live in these areas, but we excluded them from our studies and confined our vegetation analysis to the limits of the reserve, which was surrounded by a barbed-wire fence.

We made canopy maps of the home ranges of the two troops we studied by direct observation of the trees from the ground. These canopy maps showed that the home range of Troop ST (the smaller area surrounded by a dotted line at the top of Fig. 1) was made up of closed-canopy forest dominated by *T. indica* and open-canopy forest dominated by *N. isoneura.* The home range of Troop DUD (the larger area surrounded by a dotted line in Fig. 1) was made up of open-canopy forest dominated by *T. indica, N. isoneura,* or *A. rovumae* and also included a large section of brush-and-scrub habitat. Table I shows the percentage of cover of the canopy in each of these regions: Troop ST's closed canopy, Troop ST's open canopy, Troop DUD's open canopy, and Troop DUD's brush and scrub. Note that Troop ST's home range did not include any brush-and-scrub habitat; Troop DUD's home range included no closed-canopy forest.

Table I. Total Canopy Cover

	Closed canopy	Open canopy	Brush and scrub
Troop ST	88%	56%	—
Troop DUD	—	52%	12%

C. Statistical Vegetation Analysis—Ordination

We distinguished the four vegetation types (or habitats) described above using direct observation on our daily excursions into the forest. The observations were based on the following aspects of the forest: physical structure of the plants, light availability, spatial positioning of the plants, and canopy cover. In addition, we gathered numerical data on the plant species present in the forest for analysis by ordination.

A 50-m grid was superimposed on the home ranges of the two troops studied most intensely. The subdesert part of the reserve was also marked with a grid. At each intersection on this grid a 10-m quadrat was laid out. (Inside this 10-m quadrat a 4-m quadrat was marked off. The data from these smaller quadrats are discussed below in Section IID on plant-species diversity.) All large (5–15 m tall) and small (3–7 m tall) tree species and the numbers of individuals of those species were recorded in each of the 10-m quadrats.

With the use of the Cornell Ecology Programs #4 and #5 (Gauch, 1971a,b; 1973), a Bray–Curtis ordination was performed on the species-abundance data (Bray and Curtis, 1957; Gauch and Whittaker, 1972; Cottam, Goff, and Whittaker, 1973). This ordination entailed using species and numbers of individuals for computing quadrat-similarity values. These values indicate how similar each quadrat was to each of the other quadrats. Quadrats 1 and 102 contained no species in common and were among the most dissimilar pairs of quadrats. All the other quadrats were arranged, or ordinated, along an axis defined by quadrats 1 and 102 as end points. (There were 102 quadrats in all.) A second ordination was done with quadrats 4 and 80. These two quadrats fell close to each other in the first ordination but were very dissimilar themselves. Using each quadrat's relative position in the first ordination as the abscissa value and its relative position in the second ordination as the ordinate value, I have constructed Fig. 2. Note in Fig. 2 that quadrats designated by direct observation as subdesert (open triangles) and brush and scrub (closed triangles) do form quite distinct clusters, separate from each other and from most of the other quadrats. The closed-canopy (closed circles) and open-canopy (open circles) quadrats, however, do not separate out from each other.

The dashed line in Fig. 2 indicates one possible environmental gradient, soil moisture, that served to separate the habitat types. The quadrats in the lower left portion of this graph were generally closer to the river (the source of the ground water) than were the quadrats found in the upper right. Another line could be drawn on this graph, extending from the upper left to the lower right. The brush-and-scrub and subdesert quadrats would be distributed along this second axis. The actual environmental factor responsible for this second gradient is not known.

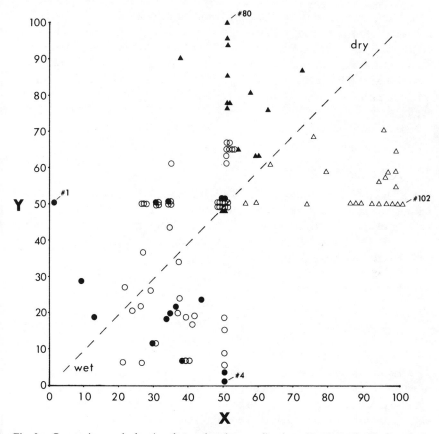

Fig. 2. Composite graph showing the results of two ordinations: X axis, ordination on plots 1 and 102; Y axis, ordination on plots 4 and 80. The dashed line indicates a moisture gradient. ●—closed-canopy forest; O—open-canopy forest; ▲—brush-and-scrub forest; △—subdesert forest.

D. Plant-Species Diversity

Another way to look for differences in habitats is to look for differences in plant-species diversity. It is possible that lemurs living in an area with many kinds of plant species may behave differently from lemurs living in an area with fewer plant species. The relative numbers of individuals of those plant species could also be important. If 75% of all plants in an area were one species, this area would be quite different from an area where that species accounted for only 10% of all individuals—even if the total number of different species in each area were the same.

We collected data on large (Level 4) and small (Level 3) trees from the 10-m plots and on shrubs (Level 2) from the smaller 4-m plots. The data consisted of species and numbers of individuals of each species. (The level classifications are from Richards, 1957, and were used for consistency with Sussman's 1972 work on *L. catta.* Level 1 is the ground level.) These data were used for the calculation (for each of the three levels in each quadrat) of *S*—number of species, *D*—density of individuals, H'—diversity, and J'—evenness.

The Shannon information index (Shannon and Weaver, 1949)

$$H' = - \sum_{i=1}^{s} p_i \log p_i$$

can be used to measure diversity when p_i equals the proportion of the total number of individuals that belong to species *i*—that is, the number of individuals of the ith species divided by the total number of individuals in the community. A larger H' indicates a greater diversity.

The evenness index (Pielou, 1969, page 233) is:

$$J' = \frac{H'_{\text{observed}}}{H'_{\text{maximum}}}$$

J' indicates the way in which the individual plants are distributed among the various species. $J' = 1.00$ would indicate that there were equal numbers of individuals in each species.

Figures 3 and 4 each show four graphs of these data. The mean values of each parameter (over all quadrats of a given vegetation type) are graphed against forest level. In Fig. 3, the three different symbols on the graphs indicate the three habitats used by the lemurs that were distinguished by direct observation (physical structure of the forest). In Fig. 4, the two symbols on each graph indicate the two habitats distinguished by the Bray–Curtis ordination (plant species). When these data were tested by analysis of variance among habitats, there was no significant difference in the upper two levels

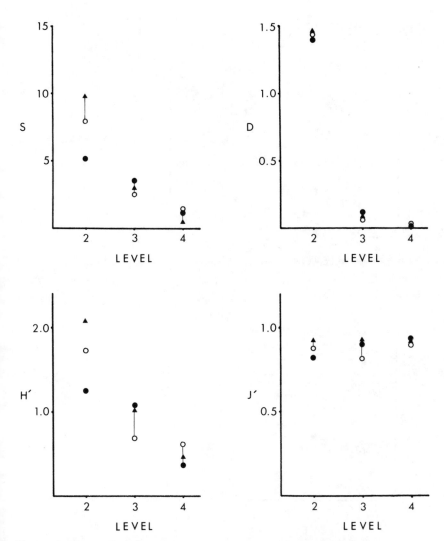

Fig. 3.   Graphs of (S) number of plant species, (D) density of individuals, (H′) diversity, and (J′) evenness in three forest levels: (2) shrub layer, (3) small trees, and (4) large trees. Habitats determined by physical structure: ●—closed-canopy forest; ○—open-canopy forest; ▲—brush-and-scrub forest. Points joined by vertical lines or touching are not significantly different.

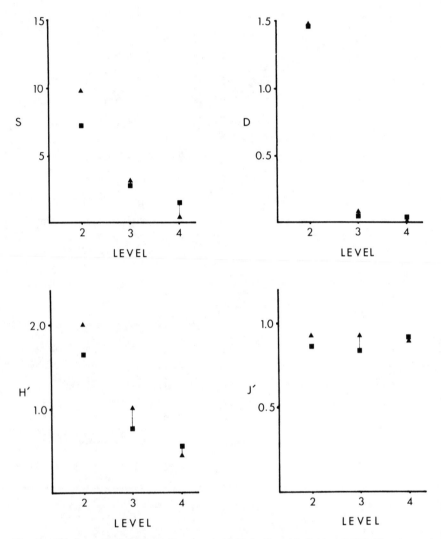

Fig. 4.   Graphs of (S) number of plant species, (D) density of individuals, (H′) diversity, and (J′) evenness in three forest levels: (2) shrub layer, (3) small trees, and (4) large trees. Habitats determined by plant species (Bray–Curtis ordination): ■—canopy forest; ▲—brush-and-scrub forest. Points joined by vertical lines or touching are not significantly different.

of the forest (Levels 3 and 4), but there was a significant difference in Level 2. The Student–Newman–Keuls test showed that in Level 2, the closed-canopy habitat was significantly different (at $p < 0.05$) from the open canopy or brush-and-scrub habitats for S (number of species) and J' (evenness), while all three habitats were different for H' (diversity). (See Fig. 3. Symbols joined by a line or touching are *not* significantly different.) In Fig. 4, the canopy habitat was significantly different from the brush-and-scrub habitat in Level 2 for S, H', and J'.

Note that the graphs of S and H' (in both Fig. 3 and Fig. 4) look very similar, while the graphs of J' are flat and very close to 1.00. This indicates that any differences in diversity are much more dependent on differences in species, while differences in density do not have much of an effect. The Level 2 diversity values indicate that brush-and-scrub and open-canopy habitats have a greater diversity than the closed-canopy habitat. Later in this paper, the use of the forest levels by the two troops of lemurs is analyzed for any behavioral differences that would correspond to these diversity differences.

III. LEMUR BEHAVIOR

A. Population Structure and Home Ranges

Censusing troops of *L. catta* is a time-consuming process. It is necessary to wait for the troop to move from one part of its home range to another. At some point in this movement, the animals generally pass from one tree to another in single file. With luck, the observer can then count the animals and determine the sex and age class of each (adult, juvenile, or infant). Two complete censuses were made of all the *L. catta* troops in the Berenty reserve (see Budnitz and Dainis, 1975). The first census of all 12 troops was made from May to December 1972. The last complete census was made from February to May 1973. The first census included 123 adults, 30 juveniles, and no infants, for a total of 153 *L. catta* in the reserve. The second census included 110 adults (down 13), 14 juveniles (just enough to replace the lost adults), and 28 infants (which corresponded nicely to the 30 juveniles of the first census). The total for the second census was 152 animals. Anne Mertl (personal communication) has reported that her census of these same animals in 1975 gave a total of 155 lemurs. Therefore, it seems reasonable to assume that, for these three years at least, the *L. catta* population remained quite stable.

For a period of several months (September–November) after the birth of the infants, some of the male lemurs in the reserve changed troops (Budnitz and Dainis, 1975). For example, in September 1972, two of the males in Troop ST began spending a great deal of time with a neighboring troop, Troop WT. Within one week, they were no longer accepted as members in their old troop. During the following breeding season (April 1973), these males were seen copulating with females in their new troop. Many other males in the reserve also changed troops in 1972, and other authors (Jolly, 1972b; Sussman, 1972) have described peripheralization of males at this same time of year in 1970. This indicates that the male exchange may be an annual occurrence. Thus, the *L. catta* population at Berenty is an interbreeding population, with gene flow made possible by the males' moving from troop to troop.

Figure 5 shows the home ranges of four of the troops in the reserve. Table II lists the number of animals in each of these troops (from the 1973 census) and the size of their home ranges in square kilometers. Note that although Troop DUD had approximately the same number of animals as Troop ST, its home range was nearly *three* times as large. Also note that although Troop BR was about the same size as Troop WT, its home range was more than *twice* as large. These data indicate that larger troops do *not* necessarily have larger home ranges. Figures 1 and 5 show that home-range size seems to be dependent on habitat. Troops whose home ranges include closed-canopy habitat and access to the river have smaller home ranges than troops that live in the scrubbier habitats and that do not have access to open water.

Struhsaker (1967) studied the home ranges of vervet monkeys (*C. aethiops*) living in different habitats. His results were very similar to ours in that he found no clear correlation between group size and home-range size. Vegetation seemed to be more important in determining the extent of the home ranges of the troops. The density of monkeys in the optimal habitat included in each group's home range was very similar for three of the four groups studied. Thus, it seems that the size of the home ranges of these

Table II. Troop Size and Area of Home Ranges

	Number of animals	Area of home range (km²)
Troop ST	17	0.081
Troop DUD	16	0.231
Troop WT	12	0.060
Troop BR	13	0.129

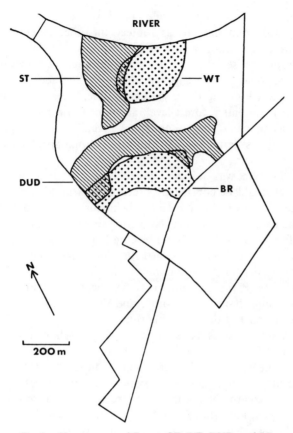

Fig. 5.   Home ranges of Troops ST, WT, DUD, and BR.

monkeys was in some way restricted by the amount of optimal habitat included.

B. Eating and Searching Behavior

Availability and distribution of food resources may be one of the most important environmental constraints on an animal. Therefore, we looked for differences in the feeding behavior of the lemurs that might correspond to the habitat and ranging behavior differences demonstrated above. Two troops of *L. catta* (Troop ST and Troop DUD) were studied for a period of four months from October 1972 to January 1973. Four adult lemurs (two males and two females) were chosen from each troop, and each of these eight animals was observed for one full day, dawn to dusk, once each month. We used adult animals as subjects in order to eliminate possible complications arising from the use of *growing* juveniles. We used males and females in order to see if there were any differences between the sexes. Every time the focus animal engaged in any feeding activity, the following data were collected: eating time, searching time, time of day, plant species eaten, part of plant eaten, forest level, and forest grid number. Eating time was defined as the time the animal spent actually eating a piece of food. Searching time was defined as the time the animal spent looking for that food.

Data collection was initiated every time the focus animal made a movement that might be construed as feeding or searching behavior. If that movement turned out to be some other behavior (grooming or fighting, for example), the data were not kept. No eating or searching during an elapsed time of 1 min indicated the end of a feeding bout. Thus, if a lemur searched for and ate leaves for 3 min, groomed for 1 min, and then searched for the same type of leaves for 2 more minutes, this behavior would be scored as two feeding bouts. If the grooming time had been less than 1 min, only one feeding bout would have been scored, and the grooming time would have been included as part of the searching time.

The first step in the analysis of this feeding behavior was to look for differences in eating time between the two troops. In other words, were the mean eating bout lengths of the two troops significantly different? Table III shows the results of an analysis of variance of the mean eating bout lengths using troops as treatment groups. Because the data were distributed log-normally, a logarithmic transformation was carried out before the analysis was done. The histogram shows the actual means. The numbers in the table are from the transformed values.

Data in Table III indicate that the means are indeed significantly different ($p < 0.001$)—Troop ST less than Troop DUD. However, there is a problem here. The numbers in the sum-of-squares column indicate the variation in the data. Notice that the Sum of Squares—Troops is a very small number compared to the Sum of Squares—Total. The ratio of these numbers is called the *coefficient of determination* and shows the percentage of the total variation in the data, which is explained by separation of the data according to troops. In this example

$$\text{Coefficient of determination} = \frac{SS_{\text{Troops}}}{SS_{\text{Total}}} = \frac{6.29}{188.53} = 0.03$$

That is, only 3% of the variation in the data is explained. In other words, *although eating-bout lengths were significantly different for these two troops, the variation explained by this way of separating the data is very small.*

Table III. Analysis of Variance Eating Time

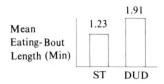

Source	Degrees of freedom	Sum of squares [log (min)]	Mean square	F	p
Total	701	188.53			
Troops	1	6.29	6.29	24.169	<0.001
Residual	700	182.24	0.26		

Coefficient of determination = 0.03

An analysis of variance of searching-bout lengths using troops as sources shows that Troop ST had a longer mean searching time than Troop DUD. Perhaps Troop ST, living in a richer environment along the river, could afford to spend more time searching for the choice pieces of food. Perhaps Troop DUD animals had to eat what they could get and could not afford to discard less desirable food. Though the difference between troops was significant, the percentage of the variation in searching time explained by the division of the data according to troops is very small.

The next step in the analysis of the feeding behavior was to look elsewhere for sources of variability that might more fully explain the variation in the data. When the mean eating and searching times were analyzed by comparison according to habitat differences (closed canopy, open canopy, brush and scrub), months (October–November versus December–January), or sexes (females of Troop ST versus females of Troop DUD, males of Troop ST versus males of Troop DUD), the results were similar. Though there were sometimes significant differences, the coefficients of determination were small—4% or less (for complete tables see Budnitz, 1976).

It is possible that the interaction between two sources of variation may explain more of the total variability than the individual sources themselves. Attempts at two-way analysis of variance—different troops in different months or different troops in different habitats—yielded similar results. The interactions did not show any significance, and the coefficients of determination were still very small. Thus, the behavior of different troops in different months or in different habitats does not resolve the problem of high variability in the data.

C. Activity Rhythms

We analyzed several other aspects of lemur behavior to see if they showed any differences between the two troops. At 15-min intervals throughout the day (from dawn to dusk), observations were made on the activity of every lemur in view in the troop being studied. The behavior patterns noted included the following categories: moving, chasing, self-grooming, grooming another animal, scent marking, feeding, drinking, vocalizing, urinating, defecating, playing, sunning, and being inactive (included sleeping). Total activity for a given day was the sum of all observations of the first 11 categories (moving through playing). Total inactivity was the sum of all observations of the last two categories. The results of analysis-of-variance tests run on the daily means of total activity, total inactivity, and total activity minus feeding (this last test was run as a measure of "leisure" time) showed no significant difference between the means for the two troops in any of these tests.

D. Use of Forest Levels

Differences in plant-species diversity among the various habitats were most pronounced in Level 2, the shrub layer of the forest. The higher levels of the forest had lower diversity than the lower levels. The diversity in Level 2 was greatest in the brush-and-scrub habitat, least in the closed-canopy habitat. These differences in diversity might be reflected in the use of the forest levels by the lemurs. At 15-min intervals throughout the day (from dawn to dusk), observations were made on all lemurs in view in the troop being studied. The number of lemurs in each level of the forest was noted and the mean level of the troop was computed. The results of an analysis of variance of the mean level used by Troops ST and DUD showed significance at $p < 0.001$, but the coefficient of determination was only 5%.

E. Day Ranges

We compared the day ranges for the lemurs to determine if there were any differences in the distances traveled by the two troops. A map of the troop movements was drawn for each day of observation. The actual linear distance traveled by the troops (the day range) was measured from these maps and the means were compared. Troop DUD had a shorter mean day range than Troop ST ($p < 0.01$), even though its home range was three times as large (Table II). The coefficient of determination in this analysis of variance was 13%. This is the highest coefficient of determination that has been found for any of the behavioral data analyzed in this paper so far.

The day range of Troop ST was longer than that for Troop DUD. Troop ST often moved through its home range, covering the same area twice in a day. Troop DUD, on the other hand, rarely retraced its steps on a given day. This troop often traveled from one end of its home range to the other in a single day, but even though this was a distance of over 600 m, they rarely strayed far from a straight path. Thus, the actual distance traveled was less than the distance covered by the circuitous paths used by Troop ST.

During the dry season, Troop DUD depended upon succulent plants (for example, *X. perrieri*) for their water. These plants grew most abundantly in the brush-and-scrub habitat. The plants that the lemurs used for food, however, were most abundant in the open-canopy part of their home range. Therefore, the lemurs may have been forced to move back and forth between these two habitats (once every day or two) in order to fill their food and water needs. Troop ST, on the other hand, had a ready supply of water from the river. During the dry season, they would visit the river twice a day to drink. These visits could account for the retracing of steps by this troop.

F. Use of *Tamarindus indica*

The size of the home ranges and the length of the day ranges were different for Troops ST and DUD. Perhaps the distribution and use of a particular food resource might correlate with these differences in ranging behavior. One of the most important food resources for both lemur troops was *T. indica*. These large, spreading trees were used for their leaf buds, their mature leaves, occasionally their flowers, and most importantly, their fruit. The fruit are pods that contain very hard seeds surrounded by a sticky, sweet, and sour matrix. The lemurs broke open the brittle, thin covering and ate the sticky matrix, seeds and all. The seeds were not digested and simply passed through the lemurs and were excreted. It is possible, of course, that lemurs play an important part in spreading tamarind seeds. The lemurs ate ripe tamarind pods (as did their observers) and also ate dried pods that had fallen, even after the pods had been on the ground for several months. The hot, dry weather in Berenty kept many pods from rotting for long periods of time.

Table IV shows the percentage of the home ranges of each troop that

Table IV. Tamarind Canopy Cover

	Closed canopy	Open canopy	Brush and scrub
Troop St	52%	17%	—
Troop DUD	—	32%	13%

was covered by tamarind canopy. It can be seen from this table that tamarinds were most abundant in the closed-canopy habitat, where Troop ST spent most of its time (see Section IIIG). Similarly, in Troop DUD's home range, tamarinds were more abundant in the open-canopy habitat, where this troop spent most of its feeding time.

Direct measurements of fruit production (counting pods that had fallen to the ground) did not show any significant differences among the different habitats. However, when indirect measurements of fruit production were analyzed, significant differences ($p < 0.025$ or better) and high coefficients of determination (33–44%) were found. Crown size (as suggested by Struhsaker, 1974), height, and diameter at breast height each gave these results when tamarinds in different habitats were compared.

The real differences found were between the closed-canopy and the open-canopy tamarinds. Therefore, in terms of tamarinds, at least, it appears that Troop ST had a home range that was richer than the home range of Troop DUD. The size measurements show that the trees were bigger, and the canopy data show that they were more densely distributed.

Analysis of variance tests on mean lengths of feeding and searching bouts in tamarinds were run on the same treatment groups described earlier for all food species. The results were essentially the same. Where there are significant differences among means, the coefficients of determination are still very low and most of the variability in the data is left unexplained. When the feeding times on the various parts of the tamarinds (buds, leaves, or fruits) were analyzed separately, the results were again the same.

G. Feeding Behavior—Total Feeding Time

It has been shown that the home ranges of Troops ST and DUD were different in various respects. The home ranges were made up of different habitats that had different physical structure or were composed of different species of plants. The distribution of tamarind trees, a major food resource, was different in the various habitats. The availability of water was different. However, when very fine-grained measures of feeding behavior—lengths of eating and searching bouts—were analyzed for differences between the two troops, the results, though revealing significant differences, did not explain the very high degree of variability.

Perhaps a more coarse-grained look at the data would explain more of this variability, though in a much more general way. Total feeding time— that is, eating time plus searching time—for the two troops was analyzed in various ways, and the results are presented below. The first step in the analysis was to see if the data from individual animals could be pooled. If the feeding times of the individual animals were not significantly different, then it would be reasonable to pool the data to get *troop* feeding times without introducing serious bias to the analysis. Two two-way analyses of variance were carried out. In the first analysis, mean total feeding time per day for the individual lemurs of Troop ST were tested against months, and there were no significant differences among individuals. Similar results were found for individual lemurs in Troop DUD. There were no significant interactions in either of these analyses.

Two-way analysis-of-variance tests were run on mean total feeding times per day of individuals in different habitats. This time, however, the habitats were defined differently from those used in previously discussed analyses. The home ranges of the two troops were divided into *prime* and *nonprime* habitats. For Troop ST, the prime habitat corresponded to the closed-canopy habitat, while the nonprime habitat was the open-canopy habitat. For Troop DUD, however, the open-canopy habitat was classed as prime, while the brush-and-scrub habitat became the nonprime habitat. The results of these analyses showed that the lemurs had very different mean total feeding times per day in these two classes of habitats ($p < 0.001$). However, there were no significant differences among the individuals in either troop, and the interaction terms were not significant. Thus, it was safe to pool the data for individuals and to look at total feeding times for the two troops. The total time the lemurs spent feeding during our study— that is, the sum of the 16 days of observations for each troop—was 31.96 hours for Troop ST and 29.98 hours for Troop DUD. This is approximately 2 hours per day.

The final step in the analysis of the total-feeding-time data was a three-way analysis of variance using months (October–November versus December–January), habitats (prime versus nonprime), and troops (ST versus DUD) as treatment groups. Table V shows the results of this analysis. The first half of this table shows the means for the various parameters. For example, Troop ST had a mean total feeding time of 113.7 min/day in October–November in their prime habitat. In the same months, their mean total feeding time was only 15.5 min in the nonprime habitat.

The second half of Table V shows the results of the analysis of variance on the total-feeding-time data. There is a significant difference between

Table V. Analysis of Variance of Total Feeding Time per Day (Min)

		Troop ST	Troop DUD
Prime habitat	Oct.–Nov.	113.74 (88.0%)	124.26 (93.2%)
	Dec.–Jan.	101.86 (92.2%)	67.11 (73.3%)
Nonprime habitat	Oct.–Nov.	15.50 (12.0%)	8.96 (6.8%)
	Dec.–Jan.	8.61 (7.8%)	24.51 (26.7%)

N = 8 in each cell

Source	Degrees of freedom	Sum of squares	Mean square	F	p	CD^a
Total	63	185228.8				
Months	1	3643.6	3643.6	4.339	<0.05	0.02
Troops	1	220.9	220.9	0.263	ns[b]	—
Habitats	1	122071.6	122071.6	145.377	<0.001	0.66
Month–troop interaction	1	521.6	521.6	0.621	ns[b]	—
Troop–habitat interaction	1	1128.1	1128.1	1.343	ns[b]	—
Month–habitat interaction	1	6035.3	6035.3	7.188	<0.01	0.03
Month–troop–habitat interaction	1	4585.0	4585.0	5.460	<0.01	0.02
Residual	56	47022.6	839.7	—	—	—

[a] CD—coefficient of determination.
[b] ns—no significant difference.

months (p <0.05), but the coefficient of determination is very small. However, there is a very significant difference between mean total feeding times in the prime and the nonprime habitats ($p \ll 0.001$), and the coefficient of determination is 66%—by far the largest coefficient of determination found for any of the behavioral data in this study. The lemurs did indeed use their habitats differently, and a great deal of the variability of these total feeding time data is explained by this difference. The interactions between months and habitats and among months, habitats, and troops are also significant.

An inspection of the means and percentages listed in Table V gives an understanding of the results of the three-way analysis of variance. Note that in October–November, generally the end of the dry season in southern and southwestern Madagascar, Troop ST spent 88% of its total feeding time in its prime habitat (the closed-canopy part of its home range). In December–January, generally the beginning of the rainy season, they did not change their behavior very much (92.2%, an increase of only 4.2%). Troop DUD showed very similar behavior in October–November, spending 93.2% of their time in their prime habitat (open-canopy habitat, in this case). However, in December–January, their behavior changed dramatically. Troop DUD lemurs spent a full 26.7% of their feeding time in the brush-and-scrub part of their home range. The rainy season in southern Madagascar usually begins in December and ends in March. In this particular year, the rains did not really get going until April!

Troop ST drank directly from the river twice a day during December–January. Troop DUD did not have access to open water and had to rely on water in plants to supply their needs. The succulent plant that they ate most often was a round-leafed vine, *X. perrieri* (family Cucurbitaceae), which grew only in the brush-and-scrub habitat. Thus, it seems very likely that Troop DUD was forced to spend an inordinate amount of time in the less-preferred part of its home range because of the animals' need for water. One effect of the lack of rain could be seen in the females of Troop DUD that had given birth in the previous September. These females seemed thin and their fur was exceedingly scruffy. The mothers in Troop ST, on the other hand, appeared quite normal during these same months.

IV. DISCUSSION

The data analysis presented above was carried out in an attempt to correlate differences in habitats with differences in lemur feeding behavior. The vegetation analysis showed that there were indeed differences in the habitats occupied by the lemurs. When the various parts of the reserve were compared according to the physical structure of the forest, three basic habitats were easily differentiated: closed-canopy forest, open-canopy forest, and brush and scrub. These habitats could be distinguished by such factors as canopy cover, light availability, and spatial positioning of plants. Differences in plant-species diversity in the shrub layer of the reserve supported the distinction of these three habitats. Analysis of plant species could not distinguish between closed- and open-canopy habitats but did separate the canopy habitats from the brush-and-scrub habitat.

Various aspects of the feeding and ranging behavior of two troops of *L. catta* were analyzed. Analyses carried out at the troop level showed marked differences that correlated with habitat differences. Troop ST, the river troop, had a much smaller home range than Troop DUD, the troop whose home range did not border the river. Troop ST had a longer day range, often retracing its steps in one day, while Troop DUD rarely did this. Troop ST tended to spend its time a little higher in the forest than Troop DUD, reflecting differences in the structure of the canopy and perhaps differences in plant-species diversity. And finally, Troop ST responded differently from Troop DUD to the constraints of a *dry* wet season. In the unusually dry period in December and January, Troop DUD spent a greater than usual amount of time in its less-preferred habitat, while Troop ST did not alter its behavior at all.

The coarse-grained analysis outlined in the preceding paragraph tends to confirm the hypothesis originally presented by Crook and Gartlan (1966). They suggested that animals would react differently in different habitats. The animals would live in different-sized troops with varying social structure, and they would show differences in ranging and feeding behavior. At the troop level, *L. catta* does show significant differences in its ranging and feeding behavior. However, when fine-grained analysis of feeding behavior is carried out, the distinctions break down. When the eating and searching behavior of individual lemurs is compared, the significant differences between troops are obscured by a very large degree of variability. An individual lemur can behave very differently under very similar circumstances. It is entirely possible that on one day it might eat a tamarind pod slowly and carefully, while on another day it might eat the same pod within several seconds, or not eat it at all and just drop it to go and investigate what a neighboring lemur is eating.

Although analysis of variance, a very powerful statistical tool, shows significant differences in the behavior of the animals in the two troops, it also reveals the large degree of variability. The coefficients of determination were quite small in all of the data analyzed for individual animals. Thus, this fine-grained analysis does not really support the Crook and Gartlan hypotheses. In fact, it seems likely that individual feeding behavior is so variable that interhabitat differences may be important to the lemurs only in times of extreme environmental hardship. At the end of our study, when the rainy season was delayed, the behavior of Troop DUD changed, but any differences in the feeding behavior of the individuals could not be distinguished. The individual lemurs maintained their variable behavior even in this period of relative dryness.

The differences in habitats did correlate with differences in the gross behavior patterns of the lemur troops (for example, total time spent feeding in different habitats). These correlations did not carry over to the fine-grained analysis of the feeding behavior. However, it is possible that other aspects of individual behavior may have been affected directly by these differences in troop behavior. For example, it is possible that troops with smaller home ranges may contact neighboring troops more often. This might lead to an increase in intertroop aggression. This kind of social change could lead to changes in feeding behavior if time spent fighting began to take away from time usually spent feeding, especially under conditions of extreme environmental hardship. Therefore, a fine-grained analysis of certain aspects of the social behavior of these animals could well be a very profitable piece of research. However, I think a warning to behaviorists in general, and to primatologists in particular, is in order. It is extremely important to choose parameters that will stand up to different kinds of statistical analyses. Differences in feeding behavior, one of the basic types of behavior used by primatologists to compare primate species, may suggest one thing on the surface (that lemur troops use habitats differentially), but under more detailed scrutiny, these differences may suggest something else entirely (that the individual lemurs are facultative animals and that their behavior may be extremely variable).

V. ACKNOWLEDGMENTS

I want to thank Kathryn Dainis Jones for her essential help and support, the de Heaulme family for their hospitality in Berenty, and the government of the Malagasy Republic and all the officials who helped us. Others who helped at various stages: Peter H. Klopfer, Daniel Rubenstein, John Glasser, Paul Saunders, M. Armand Rakotozafy of O.R.S.T.O.M. in Madagascar, and the staff and students at the Duke University Primate Facility.

This work was supported in part by National Institute of Mental Health Fellowship #5 F01 MH46274-02 and by Duke University Biomedical Sciences Support Grant #5SO5-RR-07070-08.

VI. REFERENCES

Bray, J. R., and Curtis, J. T. (1957). An ordination of the upland forest communities of southern Wisconsin. *Ecol. Monogr.* 27:325–349.

Budnitz, N. (1976). Feeding behavior of *Lemur catta* in different habitats, Ph.D. dissertation, Duke University, Durham, N.C.

Budnitz, N., and Dainis, K. (1975). *Lemur catta:* Ecology and behavior, p. 219–235. In Tattersall, I., and Sussman, R. W. (eds.), *Lemur Biology,* Plenum, New York.

Clutton-Brock, T. H. (1973). Feeding levels and feeding sites of red colobus (*Colobus badius tephrosceles*) in the Gombe National Park. *Folia Primatol.* **19:**368–379.

Clutton-Brock, T. H. (1974). Primate social organization and ecology. *Nature, London* **250:**539–542.

Cottam, G., Goff, F. G., and Whittaker, R. H. (1973). Wisconsin comparative ordination. In Whittaker, R. H. (ed.), *Ordination and Classification of Communities,* Dr. W. Junk b.v., The Hague, pp. 195–221.

Crook, J. H. (1970). The socio-ecology of primates. In Crook, J. H. (ed.), *Social Behavior in Birds and Mammals,* Academic Press, New York, pp. 103–166.

Crook, J. H., and Gartlan, J. S. (1966). Evolution of primate societies. *Nature, London* **210:**1200–1203.

Eisenberg, J. F., Muchenhirn, N. A., and Rudran, R. (1972). The relation between ecology and social structure in primates. *Science* **176:**863–874.

Frisch, J. E. (1968). Individual behavior and intertroop variability in Japanese macaques. In Jay, P. C. (ed.), *Primates: Studies in Adaptation and Variability,* Holt, Rinehart, and Winston, New York, pp. 243–252.

Gartlan, J. S., and Brain, C. K. (1968). Ecology and social variability in *Cercopithecus aethiops* and *C. mitis.* In Jay, P. C. (ed.), *Primates: Studies in Adaptation and Variability,* Holt, Rinehart, and Winston, New York, pp. 253–292.

Gauch, H. G., Jr. (1971a). *Cornell Ecology Program 4: Bray-Curtis Ordination,* Cornell University, Ithaca, N.Y.

Gauch, H. G., Jr. (1971b). *Cornell Ecology Program 5: Resemblance Matrix,* Cornell University, Ithaca, N.Y.

Gauch, H. G., Jr. (1973). The Cornell ecology program series. *Bull. Ecol. Soc. Am.* **54:**10–11.

Gauch, H. G., Jr., and Whittaker, R. H. (1972). Comparison of ordination techniques. *Ecology* **53:**868–875.

Jolly, A. (1966). *Lemur Behavior,* University of Chicago Press, Chicago.

Jolly, A. (1972a). *The Evolution of Primate Behavior,* Macmillan, New York.

Jolly, A. (1972b). Troop continuity and troop spacing in *Propithecus verreauxi* and *Lemur catta* at Berenty (Madagascar). *Folia Primatol.* **17:**335–362.

Pielou, E. C. (1969). *An Introduction to Mathematical Ecology,* Wiley-Interscience, New York.

Richards, P. W. (1957). *The Tropical Rain Forest,* 4th ed., Cambridge University Press, Cambridge.

Shannon, C. E., and Weaver, W. (1949). *The Mathematical Theory of Communication,* University of Illinois Press, Urbana.

Struhsaker, T. T. (1967). Ecology of vervet monkeys (*Cercopithecus aethiops*) in the Masai-Amboseli Game Reserve, Kenya. *Ecology* **48:**891–904.

Struhsaker, T. T. (1969). Correlates of ecology and social organization among African cercopithecines. *Folia Primatol.* **11:**80–118.

Struhsaker, T. T. (1974). Correlates of ranging behavior in a group of red colobus monkeys (*Colobus badius tephrosceles*). *Am. Zool.* **14:**177–184.

Sussman, R. W. (1972). An ecological study of two Madagascan primates: *Lemur fulvus rufus* Audebert and *Lemur catta* Linnaeus, Ph.D. dissertation, Duke University, Durham, N.C.

Tattersall, I., and Sussman, R. W. (1975). Notes on topography, climate, and vegetation of Madagascar. In Tattersall, I. and Sussman, R. W. (eds.), *Lemur Biology,* Plenum, New York, pp. 13–21.

Chapter 6

STATUS AND HIERARCHY IN NONHUMAN PRIMATE SOCIETIES

Ted D. Wade

Department of Psychiatry
University of Colorado Medical Center
Denver, Colorado 80262

In recent years, two authors (Gartlan, 1968; Rowell, 1974) have examined and found wanting the application of concepts of social dominance to the behavior of, primarily, nonhuman primates. These criticisms, I believe, fulfill the purpose of ridding us of an outmoded unitary concept, but they also suffer from a degree of unclarity and excessive zeal, brushing aside many well-established, puzzling, and interesting findings. While preparing a reply to these criticisms, I found two ideas to be of underlying importance. One of these is that discussions of social dominance must guard against muddling of conceptual levels. The other is that certain time-honored approaches to the study of dominance attempt to dispense with variability when that variability is itself likely to be our clue to any further understanding. These and other, minor, novelties I hope will make this paper more than just a reply to a critique.

I. DEFINITIONS

Since many papers, including the two cited above, fail to consistently distinguish even such radically different concepts as *dominance* and *hierarchy*, some formal distinctions are desirable. It is important to distinguish among three different types of concepts: (1) the social behavior of an individual; (2) a social relationship, which consists of the interrelations of the social behavior of two individuals (a dyad); and (3) a social (group) structure, which consists of the interrelations among all the dyadic relationships in a social group. These concepts are on different levels—that is, they are hierarchically related—so that discussions of or evidence about one level of organization are related to, but not substitutable for, discussions of or evidence about another level.

A social-dominance relationship is a particular case of the class of complementary or asymmetric relationships (Bateson, 1972; Wade, 1977) in which the behavior of each partner toward the other is different. Various studies have used various kinds of behavior as indicators of social dominance—for example, the dominant partner may be the aggressor or the one who typically approaches in approach–retreat sequences, etc.; sometimes the behavior concerns priority of access to some preferred resource, such as food, space, or mates. In order not to emphasize unduly either of the roles (dominant, subordinate) that exist in dominance relationships, I shall, when appropriate, refer to them as *status* relationships. The type of social structure almost always studied in connection with status relationships has been the linear hierarchy. One may speak of degrees of linearity, with a "perfect" linear hierarchy being one in which all status relationships are transitive—that is, when A dominates B and B dominates C, then A also dominates C. This kind of hierarchy is not the same as the hierarchies of human churches, armies, and corporations, which have an ordering of various levels, but with persons on one level being of equal status. In this discussion, I use *hierarchy* to refer to the former type of structure, with its varying degrees of linearity.

II. A STRAW-MAN THEORY

In their critiques, Gartlan (1968) and Rowell (1974) have presented a complex (possibly even tangled) web of arguments whose thrust is that the phenomena subsumed under the term *social dominance* are at best artifacts of rather unnatural situations and at worst figments of our imagination. I shall attempt to separate these arguments and deal with each individually. The first argument that I consider was the most difficult to isolate—its expression in the dominance critiques has more of the character of a prevailing attitude than of a cohesive argument. Put explicitly, it amounts to saying that the facts we now have about "social dominance" do not conform perfectly well to a loosely defined unitary theory attributed to some early workers (Schjelderup-Ebbe, 1931; Maslow, 1936; Allee, 1952).

This contention is undoubtedly true, but the theory in question is so wide in scope and so unrealistic as to be virtually a straw man. Rowell and Gartlan did offer explanations (which I discuss in later sections) of some of the facts about status-related phenomena, but these explanations treat the phenomena as artifacts of highly stressful conditions. The real problem is that these explanations actually deal only with the least interesting facts (those that are closest to fitting the straw-man theory), and the authors then ignored the more interesting and complex phenomena that they originally cited in order to demolish the straw-man concept.

What exactly is the theory that they so criticized? Gartlan (1968) attributed to "*the* theory of social dominance" (italics mine) more or less explictly the following ideas: (1) all members of a species possess in greater or lesser degree a fixed amount of a unitary drive to dominate others; (2) societies are somehow organized into hierarchies on the basis of this drive; (3) most of an individual's behavior toward others is directly and simply related to the expression of dominance; and (4) individual adaptive advantage, both in survival and directly in reproduction, is conferred on an individual in proportion to that individual's dominance. Rowell (1974) too directed criticism to all of these propositions, although she was less explicit in identifying all of them as belonging to one theory.

Readers should be able to guess the nature of some of the arguments used against the unitary concept when they realize that it has virtually all the different attributes of the classical concept of an instinct: it is fixed for the life of an animal; every member of the species has it; it has a motivational character; it has a unitary adaptive function for the individual; a number of different behavior patterns are driven by it; and, perhaps most importantly, it is that which is not learned. One wonders how anything could be all these things and a mechanism of social organization as well.

Several of the arguments against the unitary-dominance concept are worth considering in more detail. One of these concerns the amount and variety of behavior supposed to be determined by dominance. Both Rowell (1974) and Gartlan (1968) reminded us of the widely known fact that the behavior patterns or other criteria used as evidence of dominance relationships do not always show good correlations either within a study or among different studies. For Rowell, this meant that the concept of social dominance is at best inconsistently useful as a "shorthand description" of behavior (essentially, an "intervening variable"—Tolman, 1932). A great deal of effort has gone into studying status relationships from the intervening-variable point of view, with continuing controversy (Bernstein, 1970; Richards, 1974; Syme, 1974) and no end currently in sight.

A historical analogy is appropriate here. Students of behavior have understood, following Hinde (1959), that even as intuitively unitary a phenomenon as hunger fails, by the intervening-variable argument, to be at all unitary. However, by treating hunger as a system (in the cybernetic sense) of interrelated variables and by a taking into account of its relations with other systems (e.g., thirst, thermoregulation), a great deal can be explained about those variables and why they "go together" (McFarland and Sibly, 1972). Perhaps the complex of phenomena subsumed under the traditional term *dominance* should be approached in the same way rather than being subjected to endless "tests" of its unitariness. We have at this point enough evidence to say that in particular situations, various kinds of social behavior and priority-of-access measures do intercorrelate to a surprising extent. The facts need explanation, and as I argue in later sections, the idea that they are simply artifacts is not very plausible.

Another line of evidence that Gartlan (1968) and Rowell (1974) used against the unitary-dominance concept is perhaps best described as the "instability" of dominance. In practice, they cited instances of instability at all three levels of analysis: instability in an individual's aggressiveness with age, instability of dyadic relationships with changes of time or changes of social context, and instability or intransitivity of positions in a hierarchy. Even though different levels of analysis were being used, the instances of instability referred to were all taken by the dominance critics as meaning that "dominance" is not an enduring trait of individuals. The direct conclusion they made from this is that no heritability for "dominance" exists and (therefore) that selective pressure cannot operate on it. This argument is bolstered, if you will, by Rowell's noting that there is no direct evidence for inheritance of social-dominance characteristics. This is a safe bet since we know very little, if anything, about the genetics of behavior in primates. It is similarly hard to take seriously Gartlan's argument that there is no phenotypic criterion of dominance on which selection can act because studies exist in which no single set of behavioral criteria can be used to place in a linear ranking all the animals of a troop from largest adult to smallest infant. There are a number of absurdities hidden in that idea, but the main ones are the assumption that a criterion has to be the same for all age–sex classes and the assumption that a relevant criterion does not exist because human beings have not found it.

A third type of argument used against the unitary-dominance concept by Gartlan (1968) and Rowell (1974) is that dominance relationships are learned. This was, at the time of Gartlan's article, not a new idea (Mason, 1964). The dominance critics have nevertheless made much of the now-obvious fact of learning involvement. Their purpose is apparently to argue that since learning is involved, genetic inheritance (and therefore natural or sexual selection) is not. Both authors have more than once made statements clearly contrasting learning and inheritance in such a way that one can make sense of them only by assuming a nature–nurture dichotomy. I mean by the latter term the theory that genes and experience are on the same logical level and are substitutable for each other, so that if one of them is found to be a necessary cause of some aspect of behavior, then the other is logically excluded as a necessary cause of that aspect.

Gartlan (1968), for instance, after stating as a problem the question of why runaway sexual selection for male dominance has not occurred, then said, "Evidence indicates that, *on the contrary*, learning plays a significant part in the assumption of social roles [he refers to dominance roles, among others], and that genetic influences are minimal" (p. 115, italics mine). The adjectives *significant* and *minimal* make it uncertain whether a dichotomy is meant, but if the argument is to make sense, then one has to assume that a "minimal" genetic influence cannot be acted upon by selection. Gartlan also more clearly stated a dichotomy: "The comparative rarity of linear hierarchies in wild populations and the more common finding of triangular relationships, reversals, and 'central hierarchies' is difficult to explain in terms of inherited tendencies or a unitary structuring mechanism, but more easily explicable as a result of learning in the dyadic situation" (p. 93). This quotation is also a good example of how readily Gartlan has mixed statements on the individual, dyadic, and group levels.

Rowell (1974) said, after a discussion of learning and dominance, "Thus far it seems that experiments and observations on dominance relationships can be explained in terms of fairly ordinary learning. . . . In particular there is nothing to suggest any genetic basis for specific dominance relationships. It hardly seems that a special concept is necessary to cover this type of behavior" (p. 138). Elsewhere, Rowell added, "Discussion of the adaptiveness of dominance is based on the assumption that it is an inherent quality of the individual. While by no means excluding a genetic component, especially in the individual's response to unusually high levels of stress, we have emphasized the importance of learning in the establishment of social relationships. There is no direct evidence of inheritance of either dominant or subordinate characteristics (p. 149). The apparent disclaimer of a dichotomy in the next-to-last sentence does little to offset the fact that genetic and learning hypotheses are being repeatedly and directly compared as if they were logically alternative explanations.

III. TRIADIC PROCESSES—ONE ALTERNATIVE

To say that studies of dominance do not fit with an overcomprehensive and simplistic formulation of early workers is only a first step. If one rejects, as I believe we must (see Sections IV and VI) the hypothesis that status-related phenomena are artifacts, what then can we make of their causation and their function? I shall make some brief remarks on function later. Any general account of causation must include some attempt to explain some of the variability in correlations of status-related behavior, the various degrees of stability and instability in dyadic status relationships, and the variety of structural features known to occur. Much about these problems may eventually be best understood, at least in some species, by a view of status-related behavior as being goal-directed and in a predominantly triadic (involving three or more animals simultaneously) rather than a dyadic (involving two animals) context. Both goal-directedness and triadic organization were proposed by Altmann (1962) for the agonistic behavior of rhesus monkeys. He noted that complex sequential and social contingencies in the individual's behavior seemed to be feedback-guided for maintaining or raising social status while minimizing any disadvantages (in aggression received, in access to commodities) in its own status. Many agonistic interactions of two animals involved a third party; for instance, animal A, in attempting to assert dominance over animal C, might "enlist" the aid of B by lip smacking, presenting, and/or glancing at B while threatening C. Other triadic patterns included passing on aggression received to a third-party scapegoat and joining in on an attack in progress. On a longer time scale, patterns such as enlisting and joining in might develop into or be characteristic of an "alliance," in which both A and B habitually affiliated with each other and mutually reinforced each other's dominance over other animals.

Since Altmann's study similar observations of a triadic basis for social-status phenomena have been extended to a number of other species (e.g., Hall and DeVore, 1965; Jay, 1965; Struhsaker, 1967; Kawai, 1965). Alliance relationships have been found to be correlated with a variety of factors, such as matrilineal relationships, habitual close affiliation, sexual consorting, and defense against others—all these in addition to the reinforcement of an alliance partner's dominance over others. Experiments on rhesus monkeys, using manipulations of group membership, have verified the connection of social status and triadic behavior (Varley and Symmes, 1966; Masserman, Wechkin, and Woolf, 1968; Vessey, 1971; Wade, 1976), but many questions of mechanisms remain unanswered. From even the little we know at this point, much of the variety and instability of status phenomena seems likely to be related to triadic processes. In focusing only on the variety and instability and not considering this type of explanation, Gartlan (1968) and Rowell (1974) were, in my opinion, throwing out the baby with the bathwater.

On the basis of extended observations of several captive groups of rhesus monkeys, I offer the following hypothesis as an example of both the complexity of the phenomena and the type of theorizing possible at the triadic (or group) level. The dynamics of alliance relationships in the groups

I observed appeared to involve a balance between individuals' attempts at upward (status) mobility by making alliances with higher-ranking animals and the countertendency of the group's structure to resist such alterations. The groups contained some half-dozen adult females each with, at times, an adult male (Wade, 1976). The first attempts at making a new alliance seemed to come almost entirely from low-ranking animals, who would solicit the male or a high-ranking female. In the case of female–male alliances, the female's motive could have been simply to be able to mate with the male and to do so without interference from other, competing, females. There were, however, a few cases in which males solicited sexual contact with low-ranking females and these females did not take advantage of the situation by allying with the male and raising their rank. The primary aim of a new-alliance attempt, then, was apparently to raise rank; rank may or may not also have carried other privileges, such as copulation.

A prospective status-raiser was usually faced with the fact that higher-ranking animals already had alliances. An important consideration was that the subordinate member, B, of a high alliance might fall in rank if its alliance partner, A, was successfully solicited away by a lower-ranking third party, C. Animal B would thus tend to resist heavily any attempt by C to solicit an alliance with A. Given a choice, C would prefer to solicit B rather than A, since the former course would be more likely to succeed. The only wrinkle in this was that if A was a male, his attractiveness seemed usually to outweigh such strategic considerations, and he, not his alliance partner, B, would be the target of solicitations from other animals. Finally, attempts by any C to ally with any B were also threatening to monkeys who ranked in between, so these would attempt to prevent C from making her move by direct attacks and/or by soliciting B against C. To sum up the strategy of a hypothetical status-raiser: solicit the lower-ranking member of a high alliance, but expect that the higher you reach, the more resistance those of intervening ranks will show; if you are in estrus, you can with more confidence attempt to ally with the very top-ranking animal, given that he is a male.

IV. IS DOMINANCE NATURAL?

A second major argument of the dominance critiques concerns how "natural" status-related phenomena may be. The argument consists of two related points: (1) that the manifestations of status on the individual, dyadic, and group levels are much more obviously like the classic picture (high agonistic behavior, rigid relationships, and linear hierarchies) in captivity than they are in wild habitats (Gartlan, 1968, p. 116; Rowell, 1974, p. 143) and (2) that, in a more absolute sense, "In wild groups hierarchies are tenuous or absent" (Rowell, 1974, p. 131), or at least "comparatively rare" (Gartlan, 1968, p. 93, quoted earlier).[1] The dominance critics have used the alleged unnaturalness of dominance as a major support for their view that status phenomena are artifacts.

The evidence actually cited by Gartlan (1968) and Rowell (1974) for point (1) supports only the position that agonistic behavior is more commonly observed in captivity. They have produced no evidence for greater rigidity of status relationships or greater linearity of hierarchies in captive as opposed to wild groups. Apparently, both authors presume that evidence suitable at one level of analysis (the individual) is just as suitable at other levels (dyad and group).

There is only one study of which I am aware in which wild and captive monkey social structures were compared with attempts to control for observer bias and the myriads of other methodological differences (see discussion of methodological problems in Rowell, 1967b) inherent in the two types of research. This study, by Rowell (1967a), found a smaller proportion of status-related ("approach–retreat") behavior in a wild population (two large groups of *Papio anubis*) than there was in a small captive group. Hierarchical structures involving adult males could not be compared because of an insufficiency of males in the caged group (see following paragraph). Rowell said (1967a p. 508), "caged females had among themselves the most markedly rigid hierarchy." No other support for this statement was provided, and there was not even a clear indication that wild females could be individually recognized. That the wild troops were considerably larger than the captive one becomes important when one considers that by chance factors alone (i.e., in the absence of some structuring mechanism), smaller groups have more linear structures (Landau, 1951a).

Rowell claimed that the wild adult males she studied showed a lack of hierarchical organization and of the patterns of alliances found by Hall and DeVore (1965); she also suggested that the difference was at least in part due to competition over tidbits furnished by park visitors in the latter study. In a companion paper, an ecological study, DeVore and Hall (1965) made no mention of such competition, and in their behavior paper they stated that it is clearly atypical, noting its presence at all only in some groups that were *not* the ones studied primarily. Even though Rowell herself made no

---

[1] Rowell confused things by attributing to Gartlan a view that hierarchies simply do not exist under really natural conditions. Such a view may be implicit, but it is nowhere stated explicitly in Gartlan's critique.

direct comparison of hierarchical and nonhierarchical natural troops, we can accept her conclusion that two different types of troops can exist, since a similar contrast was found by direct comparison (Paterson, 1973) of two troops in differing habitats. In Paterson's study, both the hierarchical and the nonhierarchical troops had ecologically impoverished environments (although they were impoverished in different ways). However, the nonhierarchical troop was the one that had apparently the most human contact (it was very near human habitations), and it also underwent a food shortage during the study period. I conclude that the difference in tendencies toward hierarchical organization, as found in troops of the common baboon, bears no simple relationship to externally imposed "stress" and/or competition for scarce resources (see also Section VI).

Point 2 of the "naturalness" argument claims that in wild troops linear hierarchies are absent or tenuous (Rowell, 1974) or else comparatively rare (Gartlan, 1968). One problem here is that even if a really wild troop had, for most or all of its members, a linear hierarchy, observational problems might very easily prevent one from confirming it. It would at any rate be unreasonable to expect a natural hierarchy to exactly fit the straw-man model that I discussed in Section II.

Given, however, reasonable interpretations of what is a "wild" troop and what is a non-"tenuous" hierarchy, does Rowell's statement about such hierarchies hold? Let us rather strictly exclude from the category, *wild*, populations of captive monkeys, free-ranging but "provisioned" monkeys, monkeys in cities, monkeys in game preserves in which they are fed by humans, and monkeys fed by experimenters in order to test status relationships, and let us define as a non-"tenuous" hierarchy only relatively linear structures that would be extremely unlikely to occur by chance in the absence of a structuring mechanism. If one uses these reasonable definitions, there are at least four studies showing nontenuous hierarchies for either all of, or large subgroups of, the members of a wild monkey troop (Jay, 1965; Simonds, 1965; Struhsaker, 1967; Poirier, 1970). If one includes hierarchies that rank different age–sex classes in a wild troop, then a study by Chalmers (1968) can be added to the list and the Poirier study underscored. Since there are really few studies of "wild" troops in which recognition of individuals (and therefore the study of hierarchies) was possible, the four or five studies cited above not only show that natural linear hierarchies do occur but also show that such structures are not in any reasonable sense "comparatively rare."

The view of Gartlan and Rowell that dominance is basically unnatural and artifactual is thus based on a confusion of levels of analysis and on ignoring much well-known evidence. At most, one can say that the evidence does not yet exclude the possibility that captivity or other types of environments may increase the probability of or the salience of status relationships and hierarchical structures.

V. DOMINANCE OR SUBORDINANCE?

Rowell's (1974) concept of *subordinance* uses behavioral and physiological evidence to argue that hierarchies consist of and are maintained by individuals who differ in a quality called *subordinance*—the tendency to show subordinate behavior. In this section, I consider only the behavioral evidence; the physiological evidence is considered later. I begin this section with a sort of philosophical perspective.

A friend who observed my concern over these issues of status claimed that the controversy over dominance and subordinance was a matter of cultural point of view—Americans believing that society's distinctions are maintained forcibly by those on top and Britons believing that the lower classes voluntarily admire and uphold the status of their betters. This idea makes somewhat better sense than Rowell's (1974) quasi-political attribution of the dominance concept to male chauvinism, because it illustrates more clearly the arbitrariness of opinions about who "causes" a social relationship. In the sense that I have been using here, a "social relationship" is not something that resides *in* either (or both) of the parties concerned, anymore than a "difference" (of anything) resides in the things that are different. The connection between individual qualities or "inherent" traits and the relationships and societies composed of individuals is a slippery question empirically[2] and is just as elusive philosophically. In both realms, this problem overlaps the nature–nurture controversy.

There are three kinds of facts that Rowell (1966, 1974) has used to suggest that it is the behavior of the subordinate partner that determines, defines, or upholds a status relationship. The first kind of facts concerned correlations of individual's overall status rankings in a captive group (*P. anubis*), with rank according to the behavior frequencies that each monkey performed dominant behavior (i.e., behavior typical of a dominant role),

---

[2] An indication of the empirical complexity of the problem is illustrated by the following (Wade, 1976). In a study of the reactions of groups of rhesus monkeys to strangers, I found significant behavioral "main effects" of (i.e., differences due to) individuals' identities, groups' identities, and aspects of the group context (i.e., the composition of the group), yet I also found significant "interaction effects" among these factors as well. I concluded that the social properties of individuals and of groups are inextricably interrelated. I mean by this that while the effects can to some extent be separately measured and studied, they cannot *occur* separately and cannot in any ultimate sense be isolated.

performed subordinate behavior, was the object of dominant behavior, or was the object of subordinate behavior. These correlations showed that behavior typically performed as part of a subordinate role was more correlated with the rank of the object than with the rank of the subject (i.e., the actor), while behavior typical of a dominant role was more correlated with the rank of the subject than with the rank of the object. One aspect of this, as Rowell emphasized, is that subordinates (animals behaving in a subordinate fashion) more carefully choose their partners with regard to the partner's rank. An aspect not pointed out by Rowell is that if a hierarchy is in part a result of differences among animals in some quality or characteristic, then these correlations suggest that the quality is dominance, not subordinance. The correlations show that animals in a subordinate role do not recognize or discriminate their own rank carefully, nor do those who dominate them do so. The opposite is true of the dominant role, for which ranks are carefully discriminated both by those performing that role and by those responding in the opposite role. Oddly enough, both Rowell (1974) and Gartlan (1968) have spoken of "confidence" as a characteristic contributing to or indicative of high dominance rank, but both these authors have also firmly denied, in other statements, the existence of any quality of "dominance" that could be a basis for individual differences in rank.

It seems clear already that what we have is a relationship of status, not attributable entirely or even largely to the partner playing either role, and that individual differences exist in predispositions to behave or to be behaved toward as if one was dominant. Therefore, it seems at least an oversimplification to say, as Rowell (1974) did in a later summary of the correlational findings, that "it was the behavior of subordinate animals in approach–retreat interaction which correlated best with rank; agonistic behavior initiated by high-ranking animals was much less well correlated" (p. 139). The correlations were not even done separately for high- or for low-ranking animals. The facts are that *all* approach–retreat patterns except one (which did not correlate with either role) showed significant correlations between rank and *dominant role*. The slight differences in these correlations for subordinate-typical and dominant-typical behavior patterns do not look like, and certainly were not demonstrated to be, *significant* differences.

An interesting explanation of these properties of status relationships can be derived from a variety of well-established learning principles. In the dominant role, a monkey behaves unpredictably as to whom it may assault next, thus putting those in a subordinate role on an intermittent reinforcement schedule for their avoidance or appeasement behavior. This intermittent schedule produces rigidity of response in all those subordinates, and this rigidity in turn is the consistent reinforcement that allows those (receiving it) in a dominant role to be variable in their behavior. It even appears as if those in a subordinate role engage in probability matching, since they respond to those dominant to them in proportion to the dominant's rank, which in turn reflects the dominant's probability of acting dominant (to anyone) to begin with.[3]

In this perspective Rowell's finding should be regarded as being potentially interesting for a wide range of disciplines (learning theory, ethology, sociology, social psychology, political science, systems theory). If one objects to the language of learning theory, one could conceivably rephrase the above analysis in terms of "confidence" of dominants and "apprehensiveness" of subordinates or in terms of high information content of dominant behavior and high redundancy in subordinate behavior. In any case, the system clearly involves or needs both partners in order to work as it does.

A second line of evidence used to support the subordinance hypothesis (Rowell, 1966) is based on the logical idea that it is the subordinate's response that usually "completes" a status transaction; that is, a dominant animal cannot chase a subordinate who will not flee. Like Rowell, I have seen this situation often enough, but I have also seen subordinates who are bluffing this way get bitten. In those cases where an aggressor is trying to raise its status by forming a coalition with a third party of higher status, a part of the goal is to make a scapegoat act subordinately. If this result is not forthcoming, sterner measures may be taken. Contingency analyses may be able to clarify our subjective interpretations of these bluffing situations, but whatever the results, these analyses do not tell in which partner a status relationship is located.

Rowell (1966) also emphasizes, as part of the "completion" argument, that subordinates can, by behaving subordinately, apparently provoke from a dominant an attack that it would not otherwise have made. I have seen this phenomenon as well, but it is an infrequent pattern in rhesus monkeys. Before concluding that this type of behavior justifies the subordinance hypothesis, one must remember that it is the dominant animals that initiate most transactions, including status transactions, as Rowell's (1966) own data showed.

Rowell based another aspect of the "completion" argument on her finding that subordinates tend to "misunderstand" the intention of dominants' "friendly" approaches. Although I think the analysis of this

---

[3] The probability matching may not turn out to be very general. Although she did not find the fact in her data, Rowell (1974) reported that frequently it is the middle-ranking monkeys of a hierarchy that are the most aggressive (e.g., Rose et al., 1971).

possibility was not as objective as it needed to be, it is intuitively reasonable. Still, the analysis indicates at most that subordinates have "control"
over some transactions on a very short-term basis. One also has to consider
(as I showed above in connection with the correlational findings) what the
dominant partner may have done previously to contribute to this style of
responding by the subordinate.

A third type of support for the subordinance hypothesis was derived
(Rowell, 1966) from the question: When a behavior pattern is characteristic
of a dominant or of a subordinate role, to what *degree* is it characteristic?
Measuring the degree by a "direction-consistency index" (the percentage of
times the behavior was directed in the "correct" direction for its role),
Rowell found greater consistency in behavior typical of the subordinate
role. The judgment "greater" seems unjustified since the overlap between
dominant and subordinate indices in the table is greater than is implied in
the text and (again) no test for the *significance* of the difference was made.
Were a difference shown, it would seem simply to reflect the confidence–
apprehensiveness dimension that I have already discussed for status
relationships.

Chance (1967) presented an apparently independent hypothesis emphasizing the role of subordinate behavior in dominance hierarchies. Using
selected aspects of the data available on nonhuman primates at that time,
he argued (1) that "attention" by subordinates to those dominant to them
was the most pervasive and universal aspect of status-related behavior; (2)
that this attention was the "binder" that held social groups or subgroups
together; and (3) that many social interactions were structured along the
direction of the attention "vector" between dominant and subordinate. It
would take us too far afield to examine in detail the difficulties in Chance's
application of these principles to particular features of primate social organization. It must suffice at this time (but see also Jolly, 1972, for a related
view) to say that each application needed additional assumptions, unsup-
plied by the theorist, to explain such things as why particular animals
directed attention to particular others, why the binding force of attention
caused some animals to "equilibrate" at one distance and some at another
distance from a dominant, and, more fundamentally, why attention was
binding or binding led to attention (whether one, or both, of these was the
hypothesis was unclear). All of these problems, I think, can ultimately be
traced to the attempt to locate the relationship in the subordinate animal.
At any rate, Chance's picture of animals as being held on the end of a taut
string attached to a dominant male has suffered from realizations of the
cardinal importance of bonds between females and their offspring in most
primate societies (e.g., Sade, 1965).

VI. STRESS, STATUS, AND LEARNING

Gartlan (1968) and Rowell (1974) have both commented on the stressing effects of captivity for wild animals and have cited some correlations (mostly in rodents) of adrenal stress responses with status competition and status rank. Each author then gave a version of a mechanism to explain the correlations and also to explain how stress, especially that of captivity and similar conditions, produces hierarchical structures. Each concluded that hierarchies in captive groups and elsewhere are merely the misfiring of a stress-response mechanism, which itself is adaptive only under more natural conditions.

Before presenting these positions, I want to mention what I believe to be the "conventional" view (e.g., Matthews, 1964) of status, stress physiology, and captivity. This view is that in species that show social status relationships, the caging together of unfamiliar animals is a situation in which such relationships have to be established. The animals that become subordinates cannot escape attacks or avoid tense status confrontations, so that they are subject to more-or-less intense physical and emotional stress. This conventional view postulates a simple mechanism to account for the data. It does not assume that status relationships and hierarchies are artifactual but allows that they are possibly exaggerated, in an explainable way, in captive groups.

Insofar as it is possible to tell, Gartlan's view is that status relationships would rarely exist in the wild but that in captive monkeys they are "forced" because of encounters over preferred space, partners, and foods. Gartlan sees the outcome (i.e., who won) of such encounters as being a matter of chance but thinks that degenerative physiological changes due to the stress of the encounters lead to an inability of an initial loser ever to alter its initial dyadic status. In other words, status as determined in captivity is both artificial and virtually irreversible. There is undoubtedly a grain of truth in this, but a wide range of reversibility of status relationships has been seen in captive monkeys; indeed, this fact was used by Gartlan and Rowell as evidence against the "unitary-dominance theory." My experience with several captive groups of rhesus monkeys is that animals who are initially of low status do not want to remain there and that they can sometimes maneuver their way up the ranks.

Gartlan's explanation was rejected by Rowell (1972, 1974), who instead saw subordinate behavior as a direct stress symptom, comparable in a sense to adrenal enlargement. In this view, stress susceptibility, as an individual trait, tends to occur in animals who also have an inherent tendency to act subordinately. The sequence of events is this: a susceptible animal enters a captive group; as a response to the stress ensuing from this event, it acts subordinately, thus eliciting dominance behavior from others. Stable relationships and a dominance hierarchy are supposed to emerge from this point because of learning. The central aspect of a stressful environment, Rowell has said, is the high frequency with which animals in it meet in direct competition over space, food, and partners. Each such confrontation provides another learning "trial" in which fixed approach–retreat (status) relationships are learned. Before discussing the problems with this theory, I must note that Rowell described it as parsimonious because it accounts for the relationship between subordinate behavior and stress in only one step, while the other theories I have mentioned need more. It does so, of course, by simply postulating a direct relation between the two factors. The reader may judge whether the theory is more parsimonious in accounting for the rest of the facts about dominance behavior, status relationships, and status hierarchies.

The first major problem with this theory is that the stressor that is supposed to elicit subordinate behavior is the competition for limited commodities. One has to compete against other animals, however, so that some social mediation through others' dominance behavior seems inescapable. The paradox, again, is traceable to the attempt to "locate" the social relationship in one of the partners. One might make the case, in a substantial modification of both Rowell's and Gartlan's hypotheses, that the individuals most weakened by the nonsocial stresses of captivity—all the strange and (therefore) noxious stimuli and the diseases—are physically unable to dominate another and so are subordinate. My experience has been, however, that sickliness of an animal upon arrival in a lab and its subsequent status rank when placed in a group are correlated poorly, at least among monkeys who can survive at all in laboratory conditions.

A second problem with the stress–subordinance hypothesis concerns the stress correlations. The few studies available on primates indicate no simple relationships but, rather, a poorly understood interaction among levels of status, different components of a nonunitary stress response, species, and social history (Hayama, 1966; Sassenrath, 1970; Leshner and Candland, 1972). One study was cited but misinterpreted by Rowell (1974). Sassenrath (1970) showed that the overresponsive (to ACTH injections) adrenals of low-ranking rhesus monkeys returned to normal levels of responsiveness when the monkeys were removed from the situation in which they were submissive. To me, this finding implies that the social conditions were responsible for the adrenal problem; the low-status animals were shown not to be in any sense *inherently* overresponsive. The latter, however, is what Rowell's hypothesis requires.

A third problem concerns whether stable tendencies to show subordinate behavior really exist to the extent that the stress–subordinance hypothesis demands. Rowell (1974), in fact, gave a counterexample without even seeming to realize it. She cited the persistence of *aggressiveness* (when removed from their natural habitat) in urban rhesus monkeys, evidenced in their domination of forest monkeys in laboratory tests (Singh, 1966). Only the persistence of *subordinance,* however, is congenial to Rowell's hypothesis. Since the urban monkeys live under intensely competitive conditions, it seems reasonable to expect that some of them would be readily subordinate if Rowell's hypothesis were true, even allowing that the very worst cases would die before capture occurred. Two other counterexamples, supplied to me by A. J. Stynes (personal communication, 1975), refer to phenomena no doubt witnessed by many laboratory researchers. In status-oriented species such as rhesus monkeys, an infant of a low- or middle-status mother is in a subordinate position for years while maturing, but, especially if it is a male, it may on maturity rather suddenly dominate a number of its previously lifelong superiors, including possibly its mother. Another example occurs when a researcher adds a strange animal to a relatively stable group (e.g., Bernstein, 1969; Wade, 1976): low-status sheep of the core group can turn into tigers defending the integrity of the group. I think that these examples illustrate that submissiveness in monkeys is less an enduring trait (learned or not) than it is a compromise, making the best of a difficult situation.

One last point should be made. Rowell (1974) claimed, as did Gartlan (1968), that it was the *rigidity* of status relationships acquired in captivity that led to linear hierarchies. As will be explained in the last section, the greater rigidity of dyadic relationships as such logically has no effect on the degree of linearity of a hierarchy (Landau, 1951a). A necessary condition for linearity is that the state of one dyadic relationship must in some way be correlated with the state of other dyadic relationships. This is another illustration that one can't substitute phenomena on one conceptual level for phenomena on another.

VII. QUESTIONS OF FUNCTION

The body of theory concerning the function of social status is extensive; no attempt at a comprehensive review will be made here. Theories have ranged from the very general—seeing status as a necessary by-product of the use of aggressive behavior to regulate interindividual distance (Kummer, 1970) or seeing status as a means of harmonious intragroup relationships (Kaufman, 1967)—to specific formulations of individual (Lack, 1966), group (Wynne-Edwards, 1962), or sexual (Crook, 1972) selection. Rowell has claimed simply that there is no question of function because status is an artifact of unusual situations that did not enter into evolutionary history. She and Gartlan have both pointed out that different species do not show the same correlations between dominance and certain (presumably) adaptive behavior, such as "policing" or mating. This statement tells us that species differ (a point made in more useful detail by Jolly, 1972); the correlations remain to be explained.

I have shown that the artifact hypothesis is unreasonable. It fails to provide an explanation for the presence, in a number of the more common monkey species, of (1) a large repertoire of component acts used most typically in expressing status; (2) highly developed abilities to learn status; and (3) likewise highly developed abilities to use agonistic and other behavior patterns in elaborate social and temporal contingencies that modulate social status in a goal-directed fashion. Just as important is the fact that status phenomena do occur naturally, contrary to claims by Gartlan and Rowell. I could also add that if status is a pathological state of captivity, why hasn't selection removed the mechanisms of status and hierarchy from those domesticated (nonprimate) species in which these phenomena are so often observed?

The dominance critics claim no evident basis for the selection of social status. In fact, their acceptance of a nature–nurture dichotomy prevents them from considering the variety of possible mechanisms by which genetic differences among animals could influence status differences, even under the assumption that status is learned. Among the possible mechanisms are aspects of physical strength and appearance, aspects of emotional temperament (e.g., Mason, 1964, suggested that species that emphasize status have a general tendency for strong reactions to stimuli), social learning skills, and the ability to predict others' behavior and thus manipulate it (Altmann, 1962).

With regard to sexual selection as a mechanism for evolution of dominance characteristics, especially among males, Rowell (1974) claimed that existing correlations of status and mating are equally well explained by differential mating according to age. Rowell, in fact, offered no direct test of the age versus the status hypotheses on any of the data in question. If age and status did turn out to be the same in correlating with mating, one would still have to determine whether or not establishment of status is the mechanism by which males of the proper age get to mate.

It has been suggested (Murton et al., 1966; Chance and Jolly, 1970) that dominance-related access to scarce food resources is adaptive to a group because subordinate members would die early in a period of shortage and leave food for others. There is experimental evidence (Southwick, 1967) of status-related consumption in times of food shortage. Such a process can be accounted for by classical individual selection only if the advantages of dominance-promoting traits tend to be balanced by disadvantages of such traits or advantages of incompatible traits. In terms of inclusive fitness (Hamilton, 1964), the disadvantage of dominance selection to subordinate individuals might be offset by their genetic relatedness to dominant animals of their troop. There is, however, exchange of adult males, often of high status, between macaque monkey troops; this exchange renders the inclusive fitness explanation problematical for those species. Rowell, in connection with her hypothesis of an inherent connection between stress responsiveness and social timidity, or "subordinance," speculated that more timid animals would be able to detect danger more rapidly and thus be of value to a troop. In spite of the weak evidence at this point, the idea of a selection for social timidity is intriguing as a counterpressure against the advantages of dominance-promoting traits.

VIII. HIERARCHY FORMATION AND MATHEMATICS

A series of papers in which Landau (1951a,b, 1965, 1968) mathematically analyzed possible mechanisms of hierarchy formation is interesting for several reasons. It shows the power of purely mathematical–logical analysis in the production and evaluation of behavioral hypotheses; it enables us to reject the idea that a linear hierarchy is likely to emerge from mere individual differences, even though that idea is intuitively appealing; and, partly as a result of the latter, it makes us realize the value of maintaining conceptual distinctions among the individual, dyadic, and group levels when dealing with social phenomena.

The work uses a hierarchy index, h, which is the observed variance of the frequency of others dominated by each animal divided by its maximum possible variance—the variance that it would have if the structure were a perfectly linear hierarchy. An index of 1.0 thus indicates a perfect hierarchy, and an index of 0.0 indicates an absence of hierarchy. The first mechanism considered was that status relationships depended in a linear probabilistic fashion on the "ability vectors" of the individuals concerned, that is, on their "individual differences" in static (inherent) tendencies to win or lose in a status contest. Several different mathematical approaches to this problem were used. The cases considered make it clear that when the subjects have ability vectors drawn from the same distribution, then social structures with a high hierarchy index (in the range of what is considered to be "linear" in the behavior literature) can be expected only for groups of very small size (about five members or less). The larger a group is, the lower h becomes. The greater the number of independent components entering into an ability vector, the more extreme is the above conclusion. Even if there is a uniform (same for all pairs) bias against reversal of the initial relationship as it was determined above, the expected degree of hierarchy in the society does not change (Landau, 1951b). This is true for all degrees of bias, including absolute irreversibility. One can see this result intuitively by noting that since the initial structure based on first encounters has a low chance probability of being linear, "freezing" this result by assuming irreversibility changes nothing.

There are two caveats that leave us at this stage with possible hierarchy mechanisms. The first is that if domination is completely determined (i.e., is made either zero or one) by *any* difference in abilities of the two animals, or if it is extremely highly correlated with abilities, then h will be either 1.0 or very close to it. Considering the case of dependence on a single ability—the most favorable for hierarchy—a correlation as high as 0.75–0.80 (Pearson biserial) between the individuals' abilities and their probability of dominating others would lead to only a marginally hierarchical society with 6 members and to definitely nonhierarchical societies as the size of the group approached and passed 10. Lower and more realistic correlations of "ability" and success (the highest found by Collias, 1943, in hens was with comb size, $r = 0.59$) are less likely to lead to hierarchy.

Possibly the purest experimental example of the unlikelihood that a hierarchy would emerge from simple individual differences is a study (Howard, 1955) of grain beetles, fighting in staged-pair encounters. In the first set of encounters of a group of 10 beetles, the hierarchy index was $h = 0.54$. This value of h is considerably below the range corresponding to the usual meaning of a linear hierarchy. Such a value is also what would be expected for a correlation of about 0.60 between a single ability factor and dominance probability. We may conclude that a hypothesis of irreversible relationships based on chance results of first encounters is inadequate to explain hierarchies of more than a few animals. Rowell's (1974), and possibly Gartlan's (1968), hypotheses of hierarchy formation are of this type.

If one relaxes the assumption that the contestants are all drawn from the same population (or if the population is multimodal, which is the same thing), then the above may not hold. Unfortunately, Landau did not attempt to solve this potentially quite complex case. This possibility and other principles are illustrated in the following study. Clark and Dillon (1974) staged paired drinking competitions among squirrel monkeys. They found that males nearly always dominated females and that the males were also drinking more and habituating to the test apparatus better prior to dominance testing. Results of the paired drinking tests gave a highly linear hierarchy ($h = 0.85$) for the group of five females and five males. The pair-testing procedure precluded the operation of some social mechanisms that might exist in a social group. However, selection of subjects from two populations (male and female) of differing drinking "skill" might have determined the outcome of 23/45 intersex pair relationships and thus might have made for a much more linear structure than would otherwise have been. The inability of this experiment and others like it to differentiate between properties of social relationships and aspects of skills at performing competitive tasks (Syme, 1974) is not at issue here, since any example of one age–sex class's dominating another could have been chosen.

Landau (1951b) next considered what he termed "social factors"—in essence, any dependence of a particular encounter's outcome on the outcome of previous encounters involving either of the two parties. Here, a Markoff chain model was used to simulate effects of repeated encounters on the structural tendencies of groups. First, it was confirmed (as explained above) that the *average value* of h does not change in a society whose relationships depend only on ability vectors. Next, he showed the lack of effect of a *uniform* bias against reversal of established relationships. A *nonuniform* bias, which is a much more difficult problem, was not studied. Animals might differ in their abilities to overturn an established relationship, and this might be either correlated or uncorrelated with their ability to win initial encounters. The correlated case intuitively seems adequate to lead to high h (intuition is not the most reliable, guide, of course), while the uncorrelated case is, to me, a complete unknown. Both cases still concern static traits of individuals.

Landau's (1951b) most successful model was one in which the likelihood of dominance in a pair was probabilistically dependent on their difference in number of others dominated in the group (we can simplify and call this the difference in their status ranks). The outcomes deduced were long-range outcomes, completely independent of the initial distribution of abilities to win or of initial outcomes of dyadic contents. Interestingly, if the dependence was *linear,* so that the maximal probability of dominance occurred only in those pairs that had the maximum possible rank difference, the long-range distribution of outcomes was scarcely more hierarchical than the distribution of outcomes expected in the original single-encounter model with static abilities. A sufficient condition for absolute hierarchy was that no animal could overturn a relationship in which the other's status rank exceeded its own by two. This is, of course, a highly *nonlinear* dependence of dominance probability on rank difference, and there are no doubt a large class of nonlinear but less "extreme" functions that would work as well.

This "way" of getting a hierarchy is, as Landau (1951b) pointed out, open to many concrete interpretations. It might involve, for instance, monkeys' abilities to observe others in social encounters (a suggestion made by Altmann, 1962, and Rowell, 1974), or it might be based only on a generalization to new encounters of prior learning that one is relatively dominant (or subordinate, if you prefer). It is important to remember that the former mechanism requires a group context, while the latter could apply to "laboratory" hierarchies based on paired encounters. Since structures based on pair encounters or on free encounters in a group environment have been known to differ (e.g., King, 1965), it is possible that the two classes of mechanism do not have to lead to the same result.

The "learning-to-lose" hypothesis might be useful as an adjunct to Rowell's (1974) frequency-of-trials hypothesis for explaining hierarchy formation in captivity. Rowell (1974) cited Ginsburg and Allee (1942) as showing evidence that mice can learn to lose (or win). However, she also cited a study by Maroney and Leary (1957) in which a similar effect was *not* produced in rhesus monkeys.

In a third paper, Landau (1965) introduced new mathematics to describe cases in which a group was built up by the addition of one member at a time. Such a situation was described as if there were a bias against new members. Experimental work by Bernstein (1969, 1971) has shown this bias in several monkey species. A uniform bias against winning over any old members leads readily enough to hierarchy even in small groups. If this bias increases with the size of the old group, the rate at which h grows with increasing group size is faster, while if in addition a bias based on the individual rank of each old member is present, the increase in h is faster still. Thus, if Bernstein's work was not enough, abstract theory also provides experimenters with a method of encouraging, or avoiding encouragement of, hierarchy formation.

I suggested earlier that in status-oriented monkey species alliances between animals are probably an integral part of the status system. Landau's work did not attempt any formal analogues of alliances; such analogues can, however, be used to explain linear hierarchy formation in a parsimonious way. Assume that because of their mutual support of each other in agonistic encounters, the members of a dyadic alliance will eventually both dominate the same set of other members of their social group. This means, in effect, that they hold consecutive rank positions (e.g., they are Number 1 and Number 2, or Number 6 and Number 7, etc.) in the status hierarchy of their group. Further assume that in an alliance, one of the members dominates the other and that defensive alliances are relatively transient, if they occur. A consequence of the latter is that bottom-ranking animals, with no one to dominate in an offensive alliance, will not make alliances. Now, consider the problem of accounting for a linear hierarchy of 5 animals in which there are $5 \times 4/2 = 10$ status relationships. One need only determine or assume the composition of 2 different dyadic alliances and 1 dominance relationship between those alliances, and, by the rules stated above, all 10 dyadic status relationships have been determined. To account for the 21 relationships in a group of 7 animals, only 3 dyadic alliances and 2 interalliance dominance relationships need to be known or explained. Accounting for the 28 relationships in a group of 8 requires only that one additionally know the direction of the dyadic relationship in the 2 bottom-ranking (nonallied) animals.

IX. SUMMARY

Valid consideration of social-dominance concepts requires that one keep in mind the distinctions among the social behavior of individuals, dyadic social relationships, and social structures composed of such relationships. Rowell and Gartlan, arguing against the use of any concept of social dominance, have assumed that the only feasible way to conceptualize it is as a rigid individual trait of a unitary motivational nature. This is an undesirable assumption because it ensures that the data on dominance will fail to fit the restrictive concept, while it limits consideration of very interesting status-related phenomena on conceptual levels other than the individual–motivational. Social status will in fact probably be best understood as a dynamic, goal-directed process involving two or more animals at a time. The dominance critics mistakenly infer that since dominance relationships involve learning, a dominance trait could have no heritability and could not, therefore, have naturally evolved. Such an inference depends upon accepting an unrealistic nature–nurture dichotomy. The critics' attempt to bolster this idea of dominance as an artifact by claiming that status phenomena do not occur naturally is, I have shown, based on a combination of ignoring pertinent evidence and confusing levels of analysis.

Rowell hypothesized that status relationships are initiated and maintained by those playing the role of subordinate. I argued that to "locate" primary responsibility for a dyadic relationship in one of the partners is epistemologically unsound and that a closer examination of Rowell's evidence supports a conclusion of necessary and simultaneous involvement of both partners.

The dominance critics also use the correlation of low social status and stress debilitation in captivity to argue for the artifactual nature of status phenomena. Their accounts of mechanisms for this artifact assume an unrealistic rigidity of status relationships in captivity. Rowell has also assumed that a positive relationship between stress susceptibility and social timidity exists as an enduring trait of individual monkeys; however, the weight of the evidence is against her. Finally, the mechanism that Rowell proposes runs into logical difficulty in the assumption that the subordinate animal "causes" a status relationship.

The various arguments that dominance or status phenomena are artifacts of stressful situations do not hold up. We are thus still left with a variety of hypotheses, at various levels, concerning the evolutionary origins of social status. A common problem is explaining what factors may counteract selection for dominance-promoting traits. Rowell's speculation about the adaptiveness, in natural circumstances, of social timidity provides a new possibility of a counterfoil to selection for social dominance.

Landau's mathematical work, because it assumes the proper distinctions among individual capabilities, dyadic relationships, and group structures, shows the insufficiency of attempting to explain a group's hierarchical structure on the basis of individuals' behavioral traits. Observational learning, learning to lose (or to win), and "establishment bias" are among the mechanisms that could logically correspond to the social connections that Landau has shown are needed to produce linear status hierarchies. I showed that systems of dyadic "alliances" may also parsimoniously account for linear hierarchies.

A useful general conclusion at this point is that calling certain behavior patterns dominance and/or subordinance behavior does not serve, by itself, as a good explanation of them but that it is a valid way of classifying a range of similar behavioral phenomena whose mechanisms and functions, on several conceptual levels, present us with interesting and important problems.

X. ACKNOWLEDGMENTS

My sincere thanks to I. Charles Kaufman and A. J. Stynes for very helpful criticisms and discussions during the preparation of this chapter. This chapter was prepared while the author was recipient of a U.S.P.H.S. Postdoctoral Fellowship, #MH 00322.

XI. REFERENCES

Allee, W. C. (1952). Dominance and hierarchy in societies of vertebrates. In *Structure et Physiologie des Societes Animales,* Centre National de la Recherche Scientifique, Paris, pp. 157–181.

Altmann, S. A. (1962). A field study of the sociobiology of the rhesus monkey, *Macaca mulatta*. *Ann. N.Y. Acad. Sci.* **102**:338–435.

Bateson, G. (1972). *Steps to an Ecology of Mind*, Ballantine, New York.

Bernstein, I. S. (1969). Introductory techniques in the formation of pigtail monkey troops. *Folia Primatol.* **10**:1–19.

Bernstein, I. S. (1970). Primate status hierarchies. In Rosenblum, L. A. (ed.), *Primate Behavior*, Vol. 1, Academic Press, New York, pp. 71–109.

Bernstein, I. S. (1971). The influence of introductory techniques on the formation of captive mangabey groups. *Primates* **12**:33–44.

Chalmers, N. R. (1968). The social behavior of free-living mangabeys in Uganda. *Folia Primatol.* **8**:263–281.

Chance, M. R. A. (1967). Attention structure as the basis of primate rank orders. *Man* **2**:503–518.

Chance, M. R. A., and Jolly, C. (1970). *Social Groups of Monkeys, Apes, and Men*, Jonathan Cape, London.

Clark, D. L., and Dillon, J. E. (1974). Social dominance relationships between previously unacquainted male and female squirrel monkeys. *Behaviour* **50**:217–231.

Collias, N. E. (1943). Statistical factors which make for success in initial encounters between hens. *Am. Nat.* **77**:519–538.

Crook, J. H. (1972). Sexual selection, dimorphism, and social organization in the primates. In Campbell, B. (ed.), *Sexual Selection and the Descent of Man 1871–1971*, Aldine, Chicago, pp. 231–281.

DeVore, I., and Hall, K. R. L. (1965). Baboon ecology. In DeVore, I. (ed.), *Primate Behavior*, Holt, Rinehart, & Winston, New York, pp. 20–52.

Gartlan, J. S. (1968). Structure and function in primate society. *Folia Primatol.* **8**:89–120.

Ginsburg, V., and Allee, W. C. (1942). Some effects of conditioning on social dominance and subordination in inbred strains of mice. *Physiol. Zool.* **15**:485–506.

Hall, K. R. L., and DeVore, I. (1965). Baboon social behavior. In DeVore, I. (ed.), *Primate Behavior*, Holt, Rinehart, & Winston, New York, pp. 53–109.

Hamilton, W. D. (1964). The genetical evolution of social behavior. I, II. *J. Theor. Biol.* **7**:1–52.

Hayama, S. (1966). Correlation between adrenal gland weight and dominance rank in caged crab-eating monkeys. *Primates* **7**:21–26.

Hinde, R. A. (1959). Unitary drives. *Anim. Behav.* **7**:130–141.

Howard, R. S. (1955). The occurrence of fighting behavior in the grain beetle *Tenebrio molitor* with the possible formation of a dominance hierarchy. *Ecology* **36**:281–285.

Jay, P. (1965). The common langur of north India. In DeVore, I. (ed.), *Primate Behavior*, Holt, Rinehart, & Winston, New York, pp. 197–249.

Jolly, A. (1972). *The Evolution of Primate Behavior*, Macmillan, New York.

Kaufman, J. H. (1967). Social relations of adult males in a free-ranging band of rhesus monkeys. In Altmann, S. A. (ed.), *Social Communication among Primates*, University Chicago Press, Chicago, pp. 73–98.

Kawai, M. (1965). On the system of ranks in a natural troop of Japanese monkeys. I, II. In Altmann, S. A. (ed.), *Japanese Monkeys*, Published by the editor at University of Alberta, Edmonton, Canada, pp. 66–104.

King, M. G. (1965). The effect of social context on dominance capacity of domestic hens. *Anim. Behav.* **13**:132–133.

Kummer, H. (1970). Spacing mechanisms in social behavior. In Eisenberg, J., and Dillon, W. (eds.), *Man and Beast: Comparative Social Behavior*, Smithsonian Institution Press, Washington, D.C., pp. 219–234.

Lack, D. (1966). *Population Studies of Birds,* Clarendon Press, Oxford.

Landau, H. G. (1951a). On dominance relations and the structure of animal societies. I: Effect of inherent characteristics. *Bull. Math. Biophys.* **13**:1–19.

Landau, H. G. (1951b). On dominance relations and the structure of animal societies. II: Some effects of possible social factors. *Bull. Math. Biophys.* **13**:245–262.

Landau, H. G. (1965). Development of structure in a society with a dominance relation when new members are added successively. *Bull. Math. Biophys.* **27**:151–160.

Landau, H. G. (1968). Models of social structure. *Bull. Math. Biophys.* **30**:215–224.

Leshner, A. I., and Candland, D. K. (1972). Endocrine effects of grouping and dominance rank in squirrel monkeys. *Physiol. Behav.* **8**:441–445.

Maroney, R., and Leary, R. (1957). A failure to condition submission in monkeys. *Psychol. Rep.* **3**:472.

Maslow, A. H. (1936). The role of dominance in the social and sexual behavior of non-human primates. I: Observations at the Vilas Park Zoo. *J. Genet. Psychol.* **48**:261–277.

Mason, W. A. (1964). Sociability and social organization in monkeys and apes. In Berkowitz, L. (ed.), *Advances in Experimental Social Psychology,* Vol. 1, Academic Press, New York, pp. 277–305.

Masserman, J. H., Wechkin, S., and Woolf, M. (1968). Social relationships and aggression in rhesus monkeys. *Arch. Genet. Psychiat.* **18**:210–213.

Matthews, L. H. (1964). Overt fighting in mammals. In Carthy, J. D., and Ebling, F. J. (eds.), *The Natural History of Aggression,* Academic Press, New York, pp. 23–32.

McFarland, D., and Sibly, R. (1972). "Unitary drives" revisited. *Anim. Behav.* **20**:548–563.

Murton, R. K., Isaacson, A. J., and Westwood, N. J. (1966). The relationships between wood-pigeons and their clover food supply and the mechanism of population control. *J. Appl. Ecol.* **3**:55–96.

Paterson, J. D. (1973). Ecologically differentiated patterns of aggressive and sexual behavior in two troops of Ugandan baboons, *Papio anubis. Am. J. Phys. Anthropol.* **38**:641–647.

Poirier, F. E. (1970). Dominance structure of the Nilgiri langur (*Presbytis johnii*) of south India. *Folia Primatol.* **12**:161–186.

Richards, S. M. (1974). The concept of dominance and methods of assessment. *Anim. Behav.* **22**:931–940.

Rose, R. M., Holaday, J. W., and Bernstein, I. S. (1971). Plasma testosterone, dominance rank, and aggressive behaviour in male rhesus monkeys. *Nature* **231**:366–368.

Rowell, T. E. (1966). Hierarchy in the organization of a captive baboon troop. *Anim. Behav.* **14**:430–443.

Rowell, T. E. (1967a). A quantitative comparison of the behavior of a wild and a caged baboon group. *Anim. Behav.* **15**:499–509.

Rowell, T. E. (1967b). Variability in the social organization of primates. In Morris, D. (ed.), *Primate Ethology,* Aldine, Chicago, pp. 219–235.

Rowell, T. E. (1972). *The Social Behaviour of Monkeys,* Penguin Books, Baltimore, Md.

Rowell, T. E. (1974). The concept of social dominance. *Behav. Biol.* **11**:131–154.

Sade, D. S. (1965). Some aspects of parent–offspring and sibling relations in a group of rhesus monkeys, with a discussion of grooming. *Am. J. Phys. Anthropol.* **23**:1–18.

Sassenrath, E. N. (1970). Increased adrenal responsiveness related to social stress in rhesus monkeys. *Horm. Behav.* **1**:283–298.

Schjelderup-Ebbe, T. (1931). Die Despotie im sozialen Leben der Vögel. *Forsch. Völker-psychol. Sozialog.* **10**:77–140.

Simonds, P. E. (1965). The bonnet macaque in south India. In DeVore, I. (ed.), *Primate Behavior,* Holt, Rinehart, & Winston, New York, pp. 175–196.

Singh, S. D. (1966). The effects of human environment on the social behavior of rhesus monkeys. *Primates* **7**:33–40.

Southwick, C. H. (1967). An experimental study of intragroup agonistic behavior in rhesus monkeys (Macaca mulatta). *Behavior* **28**:182–209.

Struhsaker, T. T. (1967). Social structure among vervet monkeys (*Cercopithecus aethiops*). *Behaviour* **29**:83–121.

Syme, G. J. (1974). Competitive orders as measures of social dominance. *Anim. Behav.* **22**:931–940.

Tolman, E. C. (1932). *Purposive Behavior in Animals and Men,* Century, New York.

Varley, M., and Symmes, D. (1966). The hierarchy of dominance in a group of macaques. *Behaviour* **27**:54–75.

Vessey, S. H. (1971). Free-ranging rhesus monkeys: Behavioural effects of removal, separation and reintroduction of group members. *Behaviour* **40**:216–227.

Wade, T. D. (1976). The effects of strangers on rhesus monkey groups. *Behaviour* **56**:194–214.

Wade, T. D. (1977). Complementarity and symmetry in social relationships of non-human primates. *Primates* **18**, in press.

Wynne-Edwards, V. C. (1962). *Animal Dispersion in Relation to Social Behaviour,* Oliver & Boyd, London.

work between parent and infant in the period between birth and weaning are not confined to the reciprocal actions of observable behavior. Transactions such as the exchange of heat, the provision of nutrient, and even activation of the infant's vestibular system by the parent must be considered as well. Sensory stimulation involving inapparent pathways, such as olfaction, may be of unexpected importance to both members at this special period in their lives. Thus, one must be alert to the possibility that a given interaction may be guided and even determined by processes that are hidden from ordinary observation. But this is not the only form of regulation that these processes carry out. Heat, milk, and even ordinary sensory stimulation, when repeatedly delivered, can have unsuspected, cumulative, long-term effects. Such tonic effects of phasic stimulation have recently begun to come under systematic study in adult animals (Wenzel and Ziegler, 1977) and have been proposed by Schleidt (1973) as regulators of long-sustained adult social relationships. The existence of such processes within the early parent–infant interaction would appear to be likely, and they may act to initiate and maintain the social bond (see Section VII). But equally important, since the infant has not developed complete physiological homeostasis, the long-term effects of repeated stimulation by the mother appears to serve physiological regulatory functions for the infant. Thus, the relationship, in the form of certain specific interactions, serves as a biological as well as a behavioral regulator for the infant. This is also true to a limited extent for the mother, particularly in the area of lactation physiology. Thus, a second kind of hidden regulation may be looked for, one by which the internal states of the members are altered by specific aspects of their relationship. Finally, given the extent and importance of such regulatory influences and the rapid maturation of the infant during this period, the development of the infant may well be shaped by the action of these regulatory processes over time. In fact, the behavior (and physiology) of the mother must also change in order for the relationship to evolve so as to allow the eventual independence and autonomy of the infant. The infant may play a role in regulating these developmental changes in its parents.

However pervasive these regulatory effects of the early social relationship may be, they must act within the genetic potential of each member and be influenced by the genotypic characteristics of both parents and infant. Indeed, some behavioral and many physiological processes are relatively stable and unaffected by the interactional processes of the relationship. In this paper, I distinguish between labile and stable characteristics, in this sense, and focus upon those that are consistently influenced by definable aspects of the parent–infant interaction. I do not deal with the early infant–infant interaction, which is likely to play an important role in species with multiple births, only because there is so little information available on the roles of this interaction in very early development.

Thus, if one looks closely at the origins of social behavior in the early development of infant mammals, one soon finds oneself in the midst of a complicated web of transactions between the infant and its parents, so that a distinction between biological and psychological processes is difficult and even unwise. These realizations serve to collapse social behavior and biological regulation in the young infant into a single area of study.

The problem is how to discover these inapparent or hidden social–biological regulatory processes and to ask the question: How is the parent (or infant) regulating the behavior and internal state of the other? Imagination and a willingness to make up and test multiple competing hypotheses are essential here since we know so little about what we are likely to find. We have to begin with some assumptions. For the mother–infant dyad, it has seemed plausible to begin with the proposition that their relationship is regulated by definable interactions involving seven major systems: thermoregulatory, nutritional, vestibular, tactile, olfactory, auditory, and visual. Stimulation delivered to these systems by the interaction may be patterned, sequenced, and rhythmic and may elicit or modify hormonal or neural states in both parties that act as intervening behavioral mechanisms. In adults, most of the attention has been given to communications involving pattern and sequence (e.g., displays, learning). In the infant, and perhaps even for the parent interacting with the infant, processes involving simpler aspects of stimulation, such as level, frequency, and rhythm, may be more relevant.

Since all these systems are involved continuously in the early mother–infant interaction, how are we to carry out an experimental analysis? Two approaches are available, one altering the sensory systems and the other, the signals. In the first, one studies the relationship after one or another system has been selectively lesioned or ablated in one member of the dyad (e.g., blind mothers or anosmic infants). If a particular aspect of the social relationship is selectively altered at a certain age, one is led to further studies of that process as a regulator of that aspect of the relationship. If no changes are observed, the system can be only partially responsible. The danger here is that techniques of sensory deafferentation may render the individual abnormal (e.g., deaf or blind) and different social behavior may result because of nonspecific effects of injury or the mobilization of compensatory central neural adjustments, rather than because this was a crucial pathway of regulation of social behavior. Thus, other approaches should be used to confirm inferences drawn from lesion studies.

The second approach is based on separation of the two members of the ongoing relationship. According to the line of reasoning developed above, this separation deprives each member of all the regulatory influences previously exerted by the other. Individual aspects or factors (e.g., warmth, tactile stimulation) can then be selectively presented in the absence of other aspects of the relationship. Or surrogates lacking one or another aspect of the intact animal can be provided to test for their capacity to reverse or prevent a given change that had been observed to follow sudden complete separation. The experimental control over individual variables that can be achieved by these analytic methods appears promising. But this approach depends on the concept that an infant's responses to separation are the result of withdrawal of regulating influences previously exerted by the mother before the separation occurred. Such a view would appear to be reasonable in the case of the responses of a fetus to removal from the uterus by Caesarean section but does not appear so reasonable when used to explain the responses of a 1-year-old child to the parents' leaving for a vacation. One might say that the reason is that the first depends on "biological" processes and the second on "psychological" ones. But is this always a useful distinction, and what do we say about situations that fall between the two examples given?

There is a transition from "biological" processes to "psychological" processes in the development of the infant's relationship with its mother. And during the early social period, roughly between birth and weaning, both kinds of processes are at work simultaneously, if in fact we can distinguish between them during this period. Thus, some of the responses shown by young infants separated from their mothers may be best understood by experiments based on the concept of their being evidence of "release" from prior regulation by the interaction with the mother. Others may appear to fit more readily into the traditional conceptual framework involving social-bond formation, psychological attachment, and separation distress. By these latter formulations, behavioral, autonomic, and hormonal responses are evoked by the mother's absence ("disruption of the social bond") and are expressed as an integrated psychophysiological response similar to more familiar emotional responses, such as fear. But this body of theory has been derived from the study of older altricial mammals (e.g., 6-month-old primates) and from species with precocial development (e.g., imprinting in ducklings). Bowlby, in his elegant books on attachment and separation (1969, 1973), has reminded us that we know comparatively little about separation in younger infants (e.g., less than 6 months for the human).

In our experiments in young rats (at an age that is developmentally less advanced than that of 6-month-old primates), there is evidence that the mother functions as a regulator of physiological and behavioral function by the action of repeated stimulation over discrete pathways and that separation appears to exert most of its effects on the infant by sudden withdrawal of these regulatory influences. When the infant is released from this external regulation, new levels and rhythms of function are revealed. This regulatory function of the mother continues to operate at an age when the young are able to survive without her (i.e., are not "biologically dependent") and have already developed clear evidence of maternal attachment (see Section IIIB1). These findings have implications for our theories about maternal separation effects, which are developed in Section VII. The importance of the mother as a regulator, even after the infant can survive on its own, may help explain why vigorous and persistent attachment behavior, on both the infant's and the mother's part, has been selected by evolutionary pressures in such a broad range of mammalian species.

In this chapter, I first describe some examples of processes by which infants and mothers regulate each others' behavior and which are 'hidden' because they involve a relatively inapparent sensory system, olfaction. Then, I describe some of the cumulative effects on the infant that we have found to result from the repeated interactions characteristic of early social relationships. Since the effects are to set and modulate levels and rhythms, I refer to them as "regulatory," and since they were unexpected and are inapparent without experimental intervention, I have called them "hidden." Although this discussion is based primarily on work with the laboratory rat, I have included work with other species, particularly primates, wherever pertinent data are available. Throughout, where separation is utilized to uncover regulatory processes, the issues raised (above) about alternative concepts are discussed and whenever possible tested experimentally. Finally, the implications of the concept of regulatory processes for separation theory, social-bond formation, and development of the infant are discussed.

III. OLFACTORY PROCESSES REGULATING THE MOTHER–INFANT INTERACTION

Once we have identified the potential systems that may underlie early social processes, what can be learned of how they are used by parent and infant? A good one to start with is olfaction, since there is evidence of previously unexpected processes in both mother and infant, utilizing this pathway, in the rat.

A. Maternal Behavior

Although the immediate onset of maternal behavior in the rat after parturition appears to be induced by the hormonal changes taking place during the last hours of pregnancy, the maintenance of maternal behavior after parturition is not hormonally based and appears to involve a novel form of olfactory adaptation (Rosenblatt, 1975).

A number of important aspects of maternal behavior (e.g., retrieving, licking, and assuming the nursing position over pups) can be induced in virgin females, and even males, after several days' continuous exposure to young. It was initially supposed by Fleming and Rosenblatt (1974) that this phenomenon might be mediated by olfactory cues presented by the infants. Much to their surprise, these investigators found that mothers made anosmic by $ZnSO_4$ nasal perfusion or bilateral olfactory bulbectomy, became maternal *more* rapidly and without the usual initial period of avoidance. Additionally, induction was found to be more rapid when adults and infants were housed in small cages, ensuring proximity and promoting continuous exposure to olfactory stimulation (Terkel and Rosenblatt, 1971). This result suggested that rat pups present aversive or strange olfactory stimuli, which inhibit the expression of maternal behavior. Enforced proximity promotes adaptation. It is not clear whether the hormonal changes of late parturition act by blocking this olfactory sensitivity or by enhancing the response of the maternal behavioral systems to the tactile, thermal, auditory, and visual aspects of the stimulus complex presented by the pups.

The hidden and unexpected process that has been revealed here is the resolution of the conflicting behavioral tendencies of attraction and avoidance. The onset of specific maternal behaviors must apparently await resolution of this conflict by adaptation on the mother's part to the olfactory novelty of the young. This important process was hidden within the observable behaviors by which maternal behavior is established in the rat.

We are now in a position to ask new questions, which would not have come to mind without these findings. Do the hormones that hasten the induction of maternal behavior act by altering olfactory thresholds? Does olfaction operate as an attractive cue to the mother after the process of adaptation has taken place? What is the mechanism of olfactory adaptation? Many such questions are readily approached experimentally, and the answers will greatly enlarge our understanding of the kinds of processes that may mediate the acquisition of maternal behavior in mammals.

It is worth noting here that olfaction may play an entirely different role in other species. In the mouse, $ZnSO_4$ denervation results in cannibalism of the newborn pups by primiparous mothers (Seegal and Denenberg, 1974), and in the goat, olfaction appears to be the basis for specific recognition and bonding of mother to her young immediately after birth (Klopfer and Gamble, 1966).

B. Infant Behavior

1. The "Pheromonal Bond"

Infant rats seek out their own mother at a distance by employing olfactory stimuli alone and tend to remain close to the home-cage nest odor during a stage in their development between 15 and 27 days. Leon and Moltz

(1972) have termed this a "pheromonal bond" and have been able to identify the origin and mechanism of formation of the olfactory cue in the gastrointestinal tract of the mother. Produced in the cecum, and called "cecotrophe" by Leon (1974), the substance labels the home-cage shavings and littermates throughout the period that it is excreted in the feces of the mother. The ages at which infants respond to cecotrophe exactly correspond to the days postpartum when this material is excreted by the mother in sufficient quantity to be effective, an example of exquisite synchrony in the early parent–infant relationship. Specific recognition of the infant's own mother at a distance is possible since the olfactory qualities of the cecotrophe are determined by the mother's diet (Leon, 1975). This means that in the wild, infants will be attracted only to nests and offspring of lactating females that have foraged equally on the same food sources, a highly restricted group. In this instance, during a specified period of early development, recognition of the mother, her attractiveness to the infants, and the means by which the infant regains proximity to her appear to be mediated by the infant's olfactory sense.

2. Nursing Behavior

Most young mammals suckle readily from artificial nipples, but not so the rat. This phenomenon has created problems for the study of early nutritional influences in the rat (Plaut, 1970). For a while, it was thought that the rat mother provided some kind of physical guidance or tactile facilitation, but the readiness with which infants suckle on anesthetized mothers casts some doubt on this notion (Hall et al., 1975).

A series of studies on infant rats with olfactory deficits suggested the possibility that the infants' olfactory systems were involved in the integration of nursing. Singh and Tobach (1975) showed that infant rats that had been bilaterally olfactory bulbectomized lost weight and were rarely observed to be nursing, and many died. Because these pups were brain-damaged, we used a technique for more discrete olfactory denervation, nasal $ZnSO_4$ perfusion, and these pups closely resembled bulbectomized littermates (Singh, Tucker, and Hofer, 1976). They lost weight, those treated early had a high mortality rate, and when tested with an anesthetized mother, they showed a profound deficit in nipple orientation and attachment. Control infants given oral $ZnSO_4$ perfusion did not show these deficits. Histologic studies showed destruction of the olfactory mucosa, degeneration of the olfactory fibers, and atrophy of the glomerular layer of the olfactory bulb, the site of the first synapses.

The most interesting implication of these studies is that there may be olfactory cues present on the mother's abdomen that are vital to the pups for nipple location, attachment, and suckling. Our next study (Hofer, Shair, and Singh, 1976) provided evidence in support of this notion. Solvent extraction and washing of the mother's abdomen with acetone, alcohol, and water markedly reduced the effectiveness with which 2-week-old infants were able to initiate suckling on their anesthetized mother. Control experiments appeared to rule out thermal and tactile factors or aversion to remaining traces of the organic solvents as responsible for the failure to suckle. Behavioral observations showed that infants spent more time investigating the mother's mouth and limbs and more time away from her after the washing procedure. But they continued to nose her ventrum initially and in fact could be observed to sweep the nose and mouth repeatedly over nipples without attaching and suckling. Working independently, Teicher and Blass (1976) found that a different technique of solvent washing is equally effective and that extracts prepared from the infants' mouths, if painted on a previously washed nipple, enable the next pup to attach. We have also noted that saliva from one infant on a nipple elicits attachment by others. But how do the pups initiate suckling on unused nipples? May not the mother herself be a source of cues for suckling? We now have evidence that olfactory cues are emitted by the mother and that this emission is under hormonal control by oxytocin and not by its action on milk letdown. These findings suggest that the source of the cue for nipple orientation and attachment may be glands in the region of the nipple, perhaps similar to Montgomery's glands in the human.

These findings have come as a surprise to me since I found it hard to believe that the 2-week-old pup with its richly innervated vibrissal, lip, and mouth areas did not utilize the obvious tactile properties of the nipple in finding the right place to nurse. This is a clear example of how our assumptions tend to influence our thinking and tend to hide alternative processes from consideration.

In the case of this role for olfaction, there is some evidence for generalization to the human. MacFarlane (1975) has shown that breast-fed human infants turn their heads more frequently toward breast pads belonging to their own mothers as compared to pads from other lactating mothers, as early as the sixth day postpartum. What role olfaction plays in individual recognition, attachment, and suckling in the human remains to be explored. The rapidly advancing knowledge about regulatory processes in animals can suggest new, specific, testable hypotheses.

IV. REGULATION OF THE INFANTS' AUTONOMIC FUNCTION BY THE MOTHERS' MILK

Surprising as some of the early social proceses regulated by olfaction may be, they fall into a familiar pattern involving the organization of observable behavior according to sensory information originating from the other member of the social interaction and received by an exteroceptive system. In this section, I describe evidence that the behavior of an *internal* physiological system is controlled by stimulation that originates from the other member of the interaction but is received by an *interoceptive* system. This is the regulation of the cardiac rate of the infant by the level of milk provided by the mother.

It would not be surprising to find that a mother with insufficient milk production caused physiological changes in her offspring. But once sufficient milk is provided to allow weight gain, it is not generally supposed that the level of physiological functions is delicately regulated by the amount of milk provided, within a range of normally occurring weight gains. In this role, the mother would function as an external physiological regulatory agent controlling the behavior of internal systems in her offspring. Such a regulatory function would be more interesting if the nutrient acted through the central and autonomic nervous systems rather than through alterations in the circulatory supply of metabolic substrate. And this is what we found.

The demonstration of this relationship depended upon separating the mother and the 2-week-old infant rat, finding that there was a 40% decrease in cardiac rate and that this, in turn, was the result of a marked reduction in sympathetic cardiac tone (Hofer and Weiner, 1971). When this reduction in rate persisted unchanged despite maintenance of normal body temperature and return of the infants to nonlactating "maternal" females, our attention turned to a nutritional mechanism. At first this might appear to be a nonspecific debilitative effect of the marked weight loss sustained by these infants, except that their hearts were capable of beating at normal rates if the animals were simply stimulated by tail pinching. Then, we found that the heart-rate decline persisted even if they were fed every 4 hours by gastric intubation so that they gained a small amount of weight over the 24 hours the mother was gone (Hofer, 1970). This showed that the low heart rates were not the result of starvation. But the mother provides milk even more frequently (every 1–2 hours) at this age. A series of systematic studies with graded amounts of feeding by stomach tube demonstrated that cardiac rate was delicately tuned to the amount of milk given within the normal range of weight gain for infant rats (Hofer, 1973c). If enough milk was given by tube to produce as much weight gain as in a group of mothered infants, then heart rates of separated infants remained at the level of the mothered infants.

Variations ordinarily occur in the amount of weight gained each day by mothered infants. If the mother is disturbed or the litter size is decreased, weight gain does not occur or is reversed. Other factors, such as reduction of litter size, seem to promote weight gain. Heart rates follow these fluctuations in the weight gain of mothered infants and illustrate the cumulative action of repeated nursing bouts resulting in long-term regulation of the infants' cardiac rate by the amount of milk supplied over time by the mother.

A series of physiological studies (Hofer and Weiner, 1975) showed that this effect of nutrient on heart rate was most likely mediated by spinal sympathetic pathways and the β-adrenergic receptors on the efferent side. Since lactose and amino acids were effective only if administered intragastrically and not if administered intravenously, the receptor–afferent mechanism appeared to originate in the gut wall. Simple gastric distension, various gastrointestinal hormones, and the afferent vagus have been ruled out, but afferent mesenteric sympathetic nerves (Sharma and Nasset, 1962) remain a possible pathway by which the brain is informed of the amount of nutrient in the gut.

The heart rates of infant rats (and humans) follow a developmental course characterized by a small rise in the first days (or months), followed by a plateau of high heart rates during mid-infancy, which gives way to a slow decline during preadolescence. The high range of heart rates during infancy are the result of an initial high sympathetic tone and the subsequent decline, the result of the gradual establishment of predominant vagal restraint (Hofer, 1974). These developmental stages and transitions had been assumed to be the result of maturation of central neural homeostatic systems, and the set point at any age had been assumed to be an intrinsic neural function, probably genetically programmed, and heavily buffered from environmental influences. In these studies, we find that the age-characteristic level of heart rate is the result of the infant's nutritional relationship with its mother and can be delicately tuned, over periods of a few hours, by variations in the amount of milk the mother provides.

Why is a vital internal function, such as regulation of heart rate, at the mercy of such an unreliable control system as the relationship with the mother? The very word *autonomic* is used to denote the *autonomous* nature of the neural system that regulates certain vital internal organ systems (e.g., sympathetic and parasympathetic cardiac control). Instead, we see it in a very nonautonomous role, dependent on extrinsic regulation by another animal. The adaptive value of this regulatory phenomenon in early life may hinge on the cardiovascular tasks involved in the absorption and circulatory transport of the relatively enormous amounts of milk consumed by infant rats. At 2 weeks of age, young rats gain between 12% and 15% of body weight per day. In order to produce these weight gains by milk infusion, 10 ml of milk must be given per day to a 25–30 g animal. This means that infant rats process 30–40% of their own body weight in milk each day. A "wide-open" cardiovascular system would appear to be appropriate for this task, allowing maximal rates of transport of nutrient from the gut and delivery to all the rapidly growing peripheral tissues. With wide-open vessels, resistance to pumping is low, and the heart must have a relatively high output and high rates of pumping to maintain normal blood pressure. Normal blood pressure must be maintained at all times in order for tissue processes to function normally.) A less wide-open system would appear to be more appropriate for processing less nutrient. With decreased nutrient in the gut, we hypothesize that there will be decreased blood-vessel diameter. Concurrently, the amount and rate of blood pumped by the heart must be reduced in order to prevent a disastrous increase in blood pressure as the pumped blood meets the narrower vessels. Hence, the heart-rate adjustments serve to maintain steady blood pressure in the face of changes in the resistance of the vascular bed appropriate to the processing of changing amounts of ingested nutrient. This is our working hypothesis, supported by some pilot evidence suggesting that blood vessels *do* constrict with reduced nutrient intake. It has the advantage of proposing an adaptive value for what at first appears to be a whimsical quirk of nature.

This unexpected regulatory phenomenon of the early social relationship in the rat was revealed by the use of mother–infant separation, followed by analytic studies based on a concept of the relationship as a regulator, rather than on the concept of attachment and social-bond formation. To have inferred that the low heart rates were a reflection of an emotional state precipitated by disruption of the pheromonal bond (see above) would have been wrong and could have obscured other processes from view.

V. INFANT BEHAVIORAL AROUSAL

In this section, I outline some of the ways by which stable, long-term characteristics of the infant rat's behavior may be regulated by certain aspects of its behavioral interaction with its mother. The kind of effects I discuss here are those that are exerted over long periods of time by the cumulative actions of the repeated episodes of stimulation delivered by the mother to the infant during their ongoing social relationship. In this sense, they represent the tonic effects of phasic stimulation. In the cases described, the relationship between the particular form of stimulation and the behavioral effect cannot be deduced from simple observation of the ongoing relationship. Only upon removal of the mother for a period of hours does the behavioral effect become evident, and only upon prevention of that behavioral consequence of separation by provision of a specific form of repeated stimulation can the regulatory process be identified. These processes appear to be slow-acting ones, requiring a period of hours for behavioral change to occur after offset or onset of the altered stimulation. And the cumulative or tonic effects of a given repeated stimulation can be quite different, even opposite to the phasic effects of the stimuli.

How are the behavioral changes, which are the results of release from such tonic regulatory influences, to be distinguished from emotional responses of the infant to separation and rupture of the attachment bond? In some cases, this distinction is difficult, but if the response to separation has a short latency, if it is attenuated by the carrying out of the maternal separation in familiar surroundings, and if the presence of familiar companions significantly reduces the response, these characteristics tend to suggest attributing the response to separation distress. If the response does not follow this pattern, however, other processes should be considered. Specific regulatory processes are suggested when certain isolated aspects of the complete mother–infant interaction prevent the appearance of certain responses in the separated infant, even if they are provided artificially in the mother's absence (e.g., milk's effect on heart rate, above). Conversely, if certain mediating pathways can be blocked, the infant's response may occur despite the mother's presence (see Section V.C). These kinds of results favor relating a given separation response to release from regulation by a specific aspect of the social relationship.

This distinction and the unusual properties of these processes will become clearer in the specific examples described below and are discussed more fully in Section VII.

A. Thermal Influences

In most altricial mammals, the parents maintain the body temperature of the young by conveying their own body heat through conduction and radiation within the sheltered nest environment they have selected or constructed. The amount of thermal input is directly related to the amount of time the parents spend in contact with the young. This is a familiar example of a physiological regulatory process originating in the parent–infant social relationship: lacking fully developed thermoregulatory capacity, the body temperature of the young is regulated by how much time the parents spend with them. But can the parents regulate other behavioral and developmental processes by regulating the body temperature of their young? A partial answer to this question has been available for a number of years in data on markedly decreased heart rate, oxygen consumption, etc., in infant rats separated from their mothers and placed in refrigerators (Fairfield 1948). But these data told us only why a mother's abandoning infants in a cold climate could be lethal; it did not deal with the effects of relatively minor changes in body temperature that might occur within variations of an ongoing social relationship.

In our studies, we found that housing 2-week-old infant rats without their mother at room temperature for 18 hours significantly reduced most behaviors that are customarily elicited by placement of the infants alone in an unfamiliar observation area (e.g., self-grooming, rises, total activity count). Only a 3°C drop in body temperature occurred on the average, and the correlation between body temperature and measures of locomotion was quite high. The decreases in body temperature that so affected behavioral activity levels were not sufficient to affect heart rate in these experiments, a fall in heart rate being generally considered to be a sensitive indicator of the point at which hypothermia has reached a biologically significant level (cooling slows sinoatrial pacemaker activity).

The effects of several days' exposure to room temperature without the mother are to cause a profound motor deficit (Hofer, 1975b); to markedly slow biological maturation of brain, internal organs, and skeleton; and to hasten death, rather than to have an adaptive value in prolonging survival (Stone, Bonnet, and Hofer, 1976). The motor deficit prevents adequate foraging for food, but the developmental retardation is not simply a consequence of decreased food intake.

For these two experiments, the comparison groups were also separated from their mothers but were provided with enough thermal input to maintain warm nest temperatures (35°C). The absence of these changes in the comparison groups demonstrates that it is not the separation from the mother *per se* but the release from her thermal input that is responsible for the changes observed in the experimental group.

We can infer from these studies that the thermal input regularly provided by the repeated periodic visits of the mother serves to maintain the level of behavioral activity of the infants and to determine the rate of biological maturation of body and brain. The physiological mechanisms are not yet established by which small changes in thermal input regulate activity levels of the 2-week-old rat. But we have evidence that this is not simply the result of sluggish and unresponsive muscles or nerves, since D-amphetamine in small doses completely overcomes the motor deficit, without reversing the hypothermia.

B. Tactile Stimulation

Using the provision of body heat (above) as a model, can we identify other aspects of the mother–infant behavioral interaction that also regulate behavioral activity? To answer this question, we studied the behavior of 2-

week-old infants separated from the mother, kept at nest temperature, and provided with normal nutrient intake by constant gastric infusion (Hofer, 1973c). At the end of an 18-hour separation, these infants were markedly *more* active than normally mothered pups when placed alone in an unfamiliar observation area.

Specific behaviors showing increased levels in separated infants observed singly or in groups were: locomotion, rearing, self-grooming, defecation/urination, and total activity count. The onset of sleep was also delayed in separated infants (Hofer, 1975a). These changes, taken together, constitute what I refer to as *increased behavioral arousal*. In the following experiments, the pattern of individual behaviors was always examined, as well as the composite arousal score.

There were no differences between fed and unfed infants, and the same effect occurred in noncannulated warm infants. Furthermore, if the infants were provided with their own mother for 24 hours after she had undergone mammary ligation, these altered levels of behavior did not develop (Hofer, 1973b). From these results, nutrient and thermal factors did not seem to be likely candidates for explaining the increased behavioral responsiveness of the separated infants.

Two hypotheses compete for explanation. The hyperactivity could be viewed in the framework of the attachment hypothesis, or it could be the result of withdrawal of regulatory processes that had been acting to reduce levels of activity while the social relationship was intact. It is difficult to rule out an attachment–separation–distress model, but several predictions of that model failed to be substantiated (Hofer, 1975a). First, the hyperactivity was slow to develop—between 4 and 8 hours after separation—a slowly developing effect similar to our previous examples of regulatory process withdrawal (see above) and not characteristic of the hyperactivity of separation distress. Second, the maternally separated infants in the above experiments were housed in the familiar home cage and with seven littermates. Housing them alone and in an unfamiliar environment failed to accentuate the hyperactivity, as predicted by the attachment hypothesis, even at 8 hours, when hyperactivity was not near its maximum. Thus, the behavioral change observed did not meet our criteria for separation distress.

What sort of stimulation could serve to reduce or inhibit behavioral arousal levels during the normal mother–infant interaction? Tactile, auditory, olfactory, and vestibular stimulation are the leading candidates, the eyes not yet being open at this age. Since the mother normally licks, noses, scratches, picks up, steps on, rubs, and lies on her infants, tactile stimulation seemed a strong possibility. In order for this possibility to be tested, stimulation was provided for approximately 15 min out of every hour, in accordance with the timing of the visits and absences of the mother at this age. Littermates were housed in the same apparatus, without stimulation, for the same time period (8 hr), and then both groups were tested for 10 min of observation in an unfamiliar plastic test box. Stimulation was provided by mild electric current (0.05 mA constant amperage, just enough to elicit activity), delivered through a grid floor in densities ranging from 5/15 sec to 2/30 sec. This caused a density-related reduction in behavioral activity of treated infants when they were tested, 40 min after the last stimulation. Levels of behavior were reduced in the same pattern and to the same degree as after normal mothering over this period. A similar, although not so powerful, effect was obtained by placement of infants in a slowly rotating drum (0.5 rpm) causing them to change position every 15 sec during the episodes of scheduled stimulation. Vestibular stimulation by rocking and periodic inversion had no such effects on the behavior of the infants.

What do these results mean? They show that when tactile stimulation, which has the immediate effect of eliciting activity, is repeated, it has the cumulative long-term effect of reducing the pups' level of behavioral reactivity. The forms of tactile stimulation available to the experimenter are qualitatively and perhaps quantitatively different from those presented by the mother to her infants. We have at least mimicked the timing as closely as possible, but the behavioral activity of the infants in these experiments could have been reduced by some entirely different process than that operating within the mother–infant interaction.

With this reservation, the results can be said to be consistent with the hypothesis that levels of behavioral arousal of the infant are regulated in part by the levels of tactile stimulation delivered by the mother. A long-term quieting effect from tactile and other forms of stimulation provided to the human infant has been described by Brackbill (1971). Korner and Thoman (1970) have pointed out how effective vestibular stimulation is in calming a crying baby. Cultures differ in how human babies are stimulated, but there appears to be a widespread intuitive understanding of the necessity for regular stimulation if fussiness in infants is to be avoided. Such processes may generalize between rats and man.

Our understanding of the nature of the observed relationship is meager as yet. One possible hypothesis is based on the concept of a nonspecific stimulation requirement in early CNS development, for example, that of Roffwarg et al. (1966). When the external stimulation ordinarily provided by the mother is suddenly withdrawn by separation, a reciprocal increase in self-generated stimulation (through behavioral hyperactivity or REM sleep) would be elicited. A homeostatic regulatory system, similar to known physiological systems, would provide the organizing principle for such a relationship. Arguing in part against this formulation are the findings that marked decreases in contact stimulation (in the isolation experiments) and marked increases in some modalities of stimulation (in the unfamiliar-environment and vestibular studies) do not appreciably affect the levels of self-generated behavioral hyperactivity. Apparently, we are dealing with a system attuned only to certain specific forms or intensities of stimulation. A second hypothesis is that some active neural processes are set in motion by specific maternal stimulation, such as those underlying habituation, satiation, or extinction. Alternately, a neurotransmitter-depletion mechanism could be invoked, such as may result from much more intense unavoidable shock in the adult rat (Weiss *et al.,* 1975).

A less familiar, recent hypothesis is the "opponent-process" theory of Solomon and Corbit (1974). This is a general hypothesis that has been applied to such widely divergent phenomena as isolation calling in pre-viously imprinted ducklings and withdrawal symptoms in heroin addicts. By this theory, powerful stimuli elicit not only phasic responses with rapid onset and offset ("a-processes") but also "b-processes," which work to balance or suppress a-processes and which are slow to build up and slow to dissipate or decay. The b-process becomes indirectly evident in the slight decline in a-process phasic responses after repetition and directly evident in the period after termination of a series of stimuli. As the a-process decays much more rapidly, the b-process shows itself, after cessation of a train of regular repeated stimuli, by a change in the measure chosen, opposite in direction to the phasic effects elicited during the preceding period of repeated stimulation.

In our data, the scrambling and locomotor activity elicited by electric-current stimulation would be evidence of the a-process, and the reduced levels of behavioral reactivity following a series of repeated stimuli would be evidence of the b-process. The implication would be that the mother stimulates central neural systems underlying both processes and thereby both elicits and inhibits infant behavior by her actions upon such "opponent" processes. That such brain systems may actually exist in the infant rat is suggested by the work of Campbell *et al.* (1969), who have described an adrenergic behavioral-arousal system developing between 12 and 15 days in the young rat. And with Mabry (Campbell and Mabry, 1973), Campbell has also provided evidence for a serotonergic inhibitory system, which opposes the adrenergic system at this age, although appearing to be the weaker of the two.

Clearly, a great deal more needs to be learned about the regulation of levels of behavioral arousal by repeated tactile stimulation in the infant rat and about the central mechanisms for this effect. In addition, we need more evidence that the tactile stimulation delivered by the mother to the infant in the course of their social relationship actually triggers and controls these same mechanisms. At present, the relationship is hypothetical, and the extent and mechanism of the regulatory effect is unknown.

C. Olfactory Stimulation

In Section III above, examples were given of how both mother and infant use olfaction to regulate their social relationship. In this section, I give some evidence that the olfactory stimulation provided by the close physical proximity of the mother exerts a long-term influence on the level of behavior exhibited by her infants. The data for this inference are in part derived from an experiment (Hofer, 1975a) similar to the one described immediately above for tactile stimulation. But instead of the supplying of tactile stimulation during the mother's absence, specific olfactory stimulation was provided by the housing of the mother on top of the infants, separated from them by two layers of fine mesh 0.5 cm apart. This procedure served to attenuate the hyperactivity of complete maternal separation (in comparison to littermates housed in a similar wire mesh box without the mother) but did not do so if the infants were made anosmic by $ZnSO_4$ nasal perfusion. These results implicate olfactory rather than auditory stimulation—capable of attenuating the development of the hyperactivity of separation—as the critical factor emitted by the mother during the experiment.

From this experiment alone, it would be difficult to choose between attachment and regulatory hypotheses. The smell of the mother could serve as an indication of her presence, preventing the mobilization of separation distress evidenced by hyperactivity. Alternatively, the provision of the mother's body scent could work like the provision of her body heat, serving to regulate neural processes mediating behavioral arousal.

In an attempt to decide between these alternatives, we designed an experiment in which other indications of the mother's presence were provided, but without the olfactory stimulation. Infants were made anosmic by nasal $ZnSO_4$ perfusion and then either replaced with their mothers for 24 hours or separated for the same time period. The two groups were then compared behaviorally in the standard manner. The anosmic infants were exposed constantly to their mothers but nevertheless showed behavioral hyperactivity indistinguishable from the separated anosmic infants on three of the four behavioral measures: quadrants crossed, self-grooming, and total activity count. Only defecation/urination was significantly lower in the mothered group. (This measure may be more determined by the mother's anogenital licking of her infants than by her olfactory stimulation of them.)

It remains possible that the infants do not recognize their mother without her scent and therefore show hyperactivity as a part of separation distress, without separation's actually having taken place. But why do infant rats not show hyperactivity as soon as the mother leaves, and what evidence do we have for the inferred emotional state of separation distress?

In the interests of economy and consistency, I favor the hypothesis that specific olfactory stimulation by the mother acts over the hours and days of the physically close early social relationship and serves to modulate and reduce the infant's behavioral arousal level, in a way similar to the tactile stimulation described above. The central mechanisms for such processes presumably involve central connections between olfactory bulb and other central systems that balance and inhibit behavioral arousal systems (e.g., Campbell et al. 1969). By this line of reasoning, removal of the olfactory stimulation, either by removal of the mother or by olfactory denervation, removes the drive to a tonic inhibitory system. With the drive absent, the balancing influence of this system subsides, and the behavioral-arousal system gradually becomes able to act unopposed. The behavioral hyperactivity that develops after separation could thus be attributed to the unopposed activity of the behavioral-arousal system.

There is evidence that odors can influence the general locomotor activity of groups of mice (Bronson, 1974) and exert long-term effects on the rate of sexual maturation in females (Vandenberg, 1969). Possibly, the mother rat exerts similar long-term developmental effects on activity levels by the influence of her odor upon the neural systems mediating behavioral arousal in her infants.

D. Vestibular Stimulation

In the course of most early parent–infant interaction, the infant receives intense episodic vestibular stimulation by being picked up and carried and by position changes during nursing, clinging, and following. For example, the infant rat nurses while upside down; at any other time, such a position elicits vigorous righting responses. The infant monkey receives much vestibular stimulation by clinging to a highly mobile, gymnastic mother.

We did not find that either rocking or being turned repeatedly upside down served to prevent or attenuate the hyperactivity of separation in young rats (Hofer, 1975a). However, Thoman and Korner (1971) have shown that vestibular stimulation of infant rats in a rotation apparatus for 10 minutes each day had both the acute effect of quieting infants and the long-term effect of increasing exploratory behavior and body weight at 20 days of age, in comparison with similarly handled and swaddled infants that were not rotated.

The most dramatic example of the impact of vestibular stimulation on early development has been reported by Mason and Berkson (1974). One of the consequences of maternal deprivation beginning in the neonatal period in rhesus monkeys is the appearance of stereotyped body rocking, reaching a peak at about 6 months of age and persisting at high levels thereafter. Similar body rocking is characteristic of human infants who are severely retarded, blind, or autistic and of some otherwise normal children (Kravitz and Boehm, 1971). Among nonhuman primates, macaques, chimpanzees, baboons, geladas, and gibbons apparently show this behavior abnormality and most often as a consequence of early maternal deprivation, whether reared with an artificial surrogate mother or not. But these artificial surrogates were not mobile. Mason and Berkson gave rhesus infants a standard terry-cloth surrogate suspended on a wire so that it could swing when jumped on by the infant and furthermore gave the surrogate independent mobility by having it moved, on the end of its wire, in a circle within the cage, at irregular intervals during most of the day. The infants spent the same amount of time clinging to this mobile surrogate as to the standard stationary one, but the mobility of the surrogate completely prevented all self-rocking in these infants. Two other self-directed behaviors, self-clasping and sucking, were *not* affected by the mobility of the surrogate, nor was locomotion or distress vocalization. Infants of mobile surrogates spent slightly less total time in contact with the surrogates, after the first three months, but spent much *more* time in rough-and-tumble play with their surrogates than infants with stationary surrogates. Even after the surrogates were permanently removed at 1 year of age, no rocking appeared in the infants previously housed with mobile surrogates. In later tests with novel environments and strange intruder animals, the monkeys raised on mobile surrogates reacted with less timidity, distress vocalization, or extremes of locomotor activity and with less self-biting and self-rocking.

The implication of these results is that vestibular stimulation of the infant by the mother in the course of their social relationship has effects on the development of certain behavior patterns. This form of stimulation, as delivered in the normal mother–infant interaction, may function to channel the development of motor behavior into other patterns than self-rocking. In the absence of vestibular stimulation and the regulatory effects it has on motor behavior, the behavior of self-rocking emerges as early as one month postnatally and becomes characteristic and habitual in the juvenile and young adult. In this example, the regulatory function of the mother's stimulation is of *qualitative* features of the infant's behavior and of the direction of development, in addition to effects upon levels of behavior as in the previous examples in this section.

VI. REGULATION OF RHYTHMIC FUNCTIONS OF THE INFANT

Rhythm is such an obvious quality of early social relationship that it is surprising to realize how little we know of its influence on the developing young. In part, our lack of knowledge is due to the prevailing view that biological clocks, as the word suggests, are autonomous regulatory mechanisms and to the implication that their maturation is exclusively determined by genes. Of course, one of the major characteristics of circadian and ultradian systems in adults is their entrainment by environmental events, including temperature change, nutrient intake, and social interactions, as well as the more familiar role of light. Entrainment, however, involves only a fine tuning or setting of the biological clocks; other characteristics such as amplitude, main direction, pattern, and period can be altered only slightly, if at all, by the various entraining stimuli.

We know very little about the extent to which events that occur during very early development may shape major characteristics of the infants' biological clocks, rather than simply acting to trigger or to set them running. The synchronous dancelike interactions of human infants and their mothers were mentioned in the introduction. It is not yet clear who is controlling whom in these play sequences, but Stern's finding (1974) that it is the infant rather than the mother who breaks eye-to-eye contact most (94%) of the time indicates how we cannot assume the infant to be merely a passive recipient of maternal influences. Stern has found that the infant does not alter overall looking rate but seems to change the *distribution* of looks during half-hour or hourly segments of time in response to cues from the mother. Similarly, the daily amount of wakefulness seems to be the same under different caretaking (Sander *et al.,* 1972) or different feeding schedules (Gaensbauer and Emde, 1973), but the distribution (e.g., day-night; prefeeding–postfeeding) is regulated by the caretaker through picking the infant up, playing, etc., or by feeding.

These studies with humans could not disclose the nature of the regulation and are only correlative. Studies in rats by Levin and Stern (1975) identified a pattern of predominantly diurnal weight gain in the young rat up to about Day 17 which depended upon (1) a mother with an intact visual system and (2) access by the mother to food during the night. Young of blinded mothers had no clear day–night periodicity in weight gain, and young of mothers fed only during the day had reversed cycles (e.g., gained more weight at night). After Day 19, normal infants shifted to solid food as the major source of nutrient and to the normal adult rhythm of nocturnal feeding, even in the absence of the mother. Blind infants shifted to a nocturnal pattern at about the same time, but only if their mothers were sighted and had a similar rhythm. As soon as the mother was removed from the cage, they reverted to an eating pattern without clear day–night differentiation. Thus, the mother is the source of the diurnal pattern of weight gain in the infant rat prior to 17 days but continues to be a major determinant of the pattern of food intake of her young after 19 days only if the young lack their own visual sensory capacity.

We have little data on long-range developmental consequences of perturbations in early rhythmic functions of the parent in the early social relationship. Sander and co-workers (1972) did find a later effect of rearing human neonates in a hospital nursery for 10 days, an experience that did not allow establishment of the normal day–night activity rhythm during that time. During Days 11–25, the direction of the effect of the first 10-day experience on activity rhythm was opposite in females to that in males. There is some evidence in animal work that increased stimulation in early development hastens the development of day–night circadian differences in adrenocortical secretion (Ader, 1969) and that prenatal handling of pregnant female rats hastens development of the infants' day–night activity cycle (Ader and Deitchman, 1970). This latter evidence supports our notion only indirectly, however, since the intermediary process has never been positively identified as originating in the parent–infant social relationship itself, rather than being a delayed result of hormonal change in the pregnant female acting transplacentally.

More recently, by using frame-by-frame microanalytic techniques, Condon and Sander (1974) have discovered an unexpected relationship between parents' speech rhythms and the rhythm of neonatal limb movements. If an infant was in an awake state and if it heard human speech, even in a foreign language, the points of change of its limb movements were found to synchronize closely with sound segments of speech. Use of a tape recorder ruled out the role of the adult in promoting synchrony, and the failure of infants to synchronize to isolated vowel or tapping sounds indicated a specificity to the natural rhythms of human speech. Close synchrony was observed as early as 2 days of age. It is worth noting that this relationship was revealed by the separation of the infant from its mother and the presentation of key stimuli artificially in her absence, as in some of the experiments in Sections IV and V.

The implications of this study are far-reaching, not only for the origins of language but also for the development of nonverbal communication and subtler aspects of human communication, such as empathy. The regulating action of the parents' speech on infant motor rhythms was hidden from ordinary observation and was revealed only by new "microscopic" techniques for behavior analysis and by analytic experiments with distorted and artificial stimuli.

What has been learned about rhythmic functions from separation studies? Only a few studies have been done, and none so far has succeeded in demonstrating a role for the rhythmicity of parental behavior *per se* as maintaining characteristics of the rhythmic function in the young. Rhythmic function in infants may be exemplified by the rhythmic state transitions between slow-wave sleep, REM (rapid-eye-movement) sleep, and the awake state recorded polygraphically. Loss of the mother in young guinea pigs, (Astic and Jouvet-Mounier, 1968), rats (Hofer, 1976), and monkeys (Reite *et al.*, 1974) results in one common finding, a reduction in the overall time spent in REM sleep. Where it has been looked for, there also occurs a shortening of the average duration of slow-wave and REM states, resulting in an increase in the frequency of state transitions. We found some evidence to suggest that this effect does not represent the shortening of some basic cycle but rather the occurrence of frequent interruptions in a stable underlying cycle. In any case, the probabilities of state transitions were not altered by a 24-hour maternal absence in 2-week-old infant rats. A persistent, strong, organizing process within the infant, continuing in the absence of the mother, was further suggested by the absence of any changes in rhythmicity or amount of nonnutritive sucking in the same infants.

Thus, complete maternal separation does not produce severe disorganization of rhythmic functioning. Studies have yet to be completed that will tell us the extent to which these changes in the rhythm of state transitions are the result of individual components of the early social relationship or of the delivery of such stimulation in rhythmic patterned sequences. It seems unlikely that the loss of REM sleep is an acute or transient emotional response to separation, since it was observed over a period of weeks in rat infants raised from 3 days of age without their mothers in an incubator and fed by constant gastric infusion (Carlier *et al.*, 1974).

The rhythmic nature of the interactions taking place during the early parent–infant relationship is perhaps the least "hidden" of the processes I have discussed, and yet it remains almost unexplored territory. It would therefore appear to be a particularly fruitful area for further basic research in the ontogeny of social relationships.

VII. IMPLICATIONS FOR INFANT DEVELOPMENT

A. Toward Understanding the Responses to Early Mother–Infant Separation

The assumption that early mother–infant separation is stressful and traumatic, the profound abnormalities in behavioral development that have been found to result from prolonged maternal deprivation (Spitz, 1945), and the recent findings that mothers are also adversely affected by separation from their infants (Leifer *et al.*, 1972) have directed concepts of separation into the framework of stress psychophysiology. This focus on the *response to separation* has distracted us from considering what may be going on *during* the *previous mother–infant relationship* and from realizing that knowledge about this relationship may help explain some of the responses to separation.

At present, the behavioral and physiological responses to mother–infant separation are generally viewed as related to attachment and to the inferred stress of disrupting such a strong social "bond." An emotional-distress response is supposed to ensue with behavioral and physiological components, as in classical psychophysiological responses to imposed threat. But I have described physiological and behavioral responses to separation that do not fit this model, and I would now like to examine the concept of attachment and the response to separation from a different point of view.

In altricial mammals, the growth and formation of attachment appears to take place slowly, throughout a rather prolonged sensitive period. Initially consisting only of primitive biological-approach tendencies (e.g., thermotaxis), the attraction becomes more and more specifically directed. Coincidentally, there develops the reaction to separation, which usually consists of behaviors characteristic of emotional distress, as if the animal were in physical pain or threatened with harm. The intensity of this separation response is often taken as a measure of the strength of attachment. These events have been difficult to explain in terms of learning theory, since attachment does not depend on standard reinforcing agents (Harlow, 1958; Scott, 1962; Bowlby, 1969, 1973). Attachment is then viewed as a primary drive, and words are used that convey the sense of an unusual process of stamping in (e.g., *imprinting*), or metaphors are used, such as the *bond,* that convey a sense of the enduring character of some social attachments and the traumatic impact of separation.

But the formation of attachment develops in the young coincidentally with the hidden regulatory processes outlined above. May not the formation of attachment and the expression of separation distress be related to these regulatory processes? The infant cat, for example, is at first attracted to heat and orients along thermal gradients to the nest and the mother. In the course of development, a gradual transition takes place, through a series of intermediate regulatory processes (e.g., olfaction, Rosenblatt, 1971), to the juvenile stage when attraction is to the unique stimulus configuration of an individual adult, and the regulatory effect is primarily psychological. Coincidentally, biological and behavioral regulation has shifted from the mother–infant dyad to a wide range of environmental events and to newly matured internal homeostatic mechanisms. This transition would appear to be the result of an interaction of rapidly maturing sensorimotor, integrative, and cognitive faculties of the developing young with the changing experiences of the evolving social relationship.

I have reviewed some of the unexpected, persistent, long-range, and cumulative effects of the episodic, repeated stimulation inherent in early social relationships. It seems probable that we have only just begun to scratch the surface of this area and that many other processes will eventually be elucidated. Thus, it may not be too speculative to suggest that one of the enduring effects of early repetitive stimulation is to establish and maintain the goal-directed system of behavior that we call *attachment*. The regulatory nature of attachment has been described by Bowlby (1969, 1973), and a theoretical model for the elicitation of distress behavior upon withdrawal of the attachment object has been put forward by Hoffman and Solomon (1974), supported by Hoffman's data on imprinting in ducklings. If we conceive of attachment as a regulatory process, established and governed by repetitive specific stimulation, then the classical separation responses of vocalization, locomotor hyperactivity, and even stereotyped abnormal behavior can be viewed as "withdrawal" or release phenomena, analogous to the withdrawal response of narcotic addicts after separation from their repeated drug injections. By this line of reasoning, the results of separation of an infant (of any species) from its mother (at any developmental age) are in part a function of the tonic or cumulative effects of the stimulation the infant has received from the interaction with its mother. To a variable degree of certainty, we can draw inferences as to the prior regulatory processes from the responses shown. But only by analytic experiments with specific stimulus configurations and rhythms can we positively identify them.

It appears that withdrawal responses can occur with different latencies. Those classically ascribed to attachment in older infants occur immediately upon separation or even upon signs that separation is imminent (Bowlby, 1969, 1973). Those occurring in younger infants and described in Sections IV, V, and VI above develop slowly in the hours after separation, as if the regulatory effects of the stimulation took time to dissipate. Clearly, different neural mechanisms may underlie withdrawal effects in different systems, just as the regulatory effects of stimulation vary among the different systems and the different sources of stimulation.

The concept of long-term regulatory effects produced by repeated discrete stimuli is not commonly used in the discussion of behavioral interactions but is, after all, the fundamental process by which neurons communicate and may be the mechanism for much of what the brain does (Kuffler and Nichols, 1976). Trains of repeated nerve impulses along the axon of one neuron impinge on another, altering its membrane characteristics so that it is more or less likely to be fired by an impulse from a third neuron. Aggregates of neurons show facilitation, recruitment, and inhibitory states as a result of repeated stimuli along certain pathways. Trains of stimuli deplete neurotransmitter stores, increase transmitter synthesis rates, and even alter the position and number of receptor sites, transsynaptically. If any of this neural traffic suddenly stops, the neurons that had previously received this repetitive input change functionally because of the unbalanced force of other regulatory inputs. Depending on the mechanisms involved, the shift to new levels and patterns of activity may take seconds, hours, or days to be completed.

Thus, we can see that basic neural processes, known to operate in relatively simple neural systems, have characteristics that are strikingly similar to the characteristics of the changes we have observed following abrupt termination of the early social relationship. Other, more complex processes, such as the induction of enzymes and the gradual buildup of hormonal levels, provide other possible mechanisms for tonic regulatory action and withdrawal responses following social separation.

It is not yet clear which aspects of the stimulation provided by the mother are critical to the formation of attachment, but the following formulation is offered. In birds that show imprinting, motion of the object or flashing of a light appears to be sufficient to establish following. In altricial animals, elicitation of approach, clinging, and following may depend on a series of stimuli in a developmental progression depending on the maturation of sensory capacities. These stimuli become gradually combined into a highly specific complex or gestalt. Stimulation from these sources appears to be reinforcing at appropriate ages, eliciting approach and maintaining proximity. This stimulation also has the long-term cumulative effect of reducing emotional distress and the behavior associated with it. Removal of the mother withdraws the regulatory effect of the repetitive stimulation on the emotional state of the young, and the classical separation distress ensues.

The advantage of this formulation is not only that it allows separation experiments to be used to learn more about the effects of parent–infant interaction but also that it permits us to seek to understand some of the slower-developing effects of maternal separation (Spitz, 1945; Reite *et al.*, 1974) as the results of loss of specific regulatory actions previously provided by the mother, rather than some form of prolonged, complex emotional response. It allows us to separate out each developmental effect of separation in relation to its specific mechanism, rather than viewing the phenomenon globally as an "emotional stress response," words that do not lead to further understanding. For example, a rat pup separated for 18 hours has low heart rates and is hyperreactive behaviorally in unfamiliar surroundings. This is not an integrated psychophysiological stress response like the threat of electric shock. Rather, I have shown how the experience of maternal separation in this instance becomes translated into physiological and behavioral changes by separate and different mechanisms. By altering specific aspects of the experience, we can produce an animal that is hyperactive and has normal heart rates or one that is normally active with low heart rates (see Sections IV and V above).

The formulation put forward in this section shifts the emphasis in studies on parental separation toward the preexisting parent–infant relationship. The concept of the regulatory effects of repeated stimulation is congruent with what is currently known about neurobiological mechanisms and allows an experimental approach to the processes underlying separation phenomena, social bonds, and the long-range developmental effects of early social relationships.

B. Developmental Outcome

Although much of this paper has been concerned with the regulation of infant behavior and physiology by the parent, I have tried to emphasize, wherever possible, that it is the parent–infant *interaction* that regulates its members. The changing characteristics and behavior of the infant set in motion and delicately control a progression of different behaviors in the parents. This shift in parental behavior toward the infant appears to be as important in determining the independence and competency of the infant as the infant's own maturational progress.

From what we know already, however, it is clear that the early social relationship contains within it the means to influence the development of a number of physiological and behavioral systems in the infant. Most likely, many more processes wait to be revealed, but we can already begin to see how an individual takes shape from his or her early social environment. Biologists have bemoaned the range of individual variability for as long as statistical tests have been applied to their data, and humanists have rejoiced in the uniqueness of the individual for a great deal longer. Not the least of this variability is to be found in the patterns of mothering, even within a given genetic strain. For a while, it looked as if we had our genes to thank for this situation, but recently geneticists have reemphasized the preponderant influence of the developmental environment on the expression of phenotype (Feldman and Lewontin, 1975), particularly in neurally based systems. Thus, heritability of a given behavioral trait can often be traced not so much to the genes but to postnatal factors presumably lying hidden within the early parent–infant social relationship.

A striking example of this sort of phenomenon has come out of recent work done in Czechoslovakia by Flandera and Novakova (1974). They divided their Wistar-derived strain of rats into two populations, one that attacked and killed mice and another that did not. When infants born to each type of mother were cross-fostered at birth, the infants developed the characteristics of the foster mother rather than those of their biological parents, even though mothers and infants were not exposed to mice until the young were first tested at 30 days of age. Watching their mothers kill mice further enhanced mouse killing in the young but could not have been the source of the original effect. Clearly, the mothers shaped the development of their offspring powerfully enough to overcome whatever genetic bias may have existed. Some genic effect was evident in the pups born to mothers who were mouse killers. These animals began to show increased mouse killing at 90 days, although they had not shown it at 30 days, after having been raised by a non-mouse-killing mother. The non-mouse-killing strain continued to show mouse killing at both 30 and 90 days after rearing by a mouse-killing mother. It seems unlikely that the mother transmitted only one specific kind of behavior, and we know of no processes that could explain this sort of effect in the absence of the opportunity for specific learning. And, in fact, Flandera and Novakova reported that the mouse-killing rats tended to be generally more irritable, with high levels of non-specific excitability. Killers have been found to consume more water and salt solution than nonkillers, to fight with each other more intensively, and to engage in more precopulatory activity. Thus, a whole dimension of their behavior has been affected, of which mouse killing is only one part.

We can hypothesize that the timing, pattern, and amount of stimulation delivered by the two kinds of mothers in this experiment differed appreciably and accounted for the different developmental outcomes of their offspring. Which aspects of the interaction, which sensory systems of the young, and at which developmental age—all are questions that should be approachable experimentally.

We also have evidence that biological function and disease susceptibility, in particular, can be modified by alterations of the early social relationship. Susceptibility to gastric erosions and mortality from a transplanted tumor have been shown to be increased in rats prematurely separated from their mothers. And Ackerman, Weiner, and I (1975) have found that if infant rats are separated from their mothers one to two weeks early, at 15 days of age, they develop an extraordinary susceptibility to gastric erosions induced by restraint stress at 30 days of age. At this age, 100% of early-separated rats developed ulcers and more than 70% bled massively from them, a lethal complication almost never seen in adults. Only 5–10% of normally reared 30-day-old rats showed ulceration at this age. By 100 days of age, normally reared rats were susceptible, but early-separated ones were still more susceptible. At 200 days, no differences persisted. Furthermore, provision of a nonlactating mother from Day 15 to Day 29 markedly reduced the incidence and severity of ulcers in 30-day-old rats, showing that the mother–infant behavioral interaction itself, irrespective of nutritional effects, can modify the course of development of susceptibility in this stress-disease model.

These two examples indicate the far-reaching developmental effects that can be demonstrated to result from the action of some of the hidden regulatory processes in early social relationships. These formative influences begin earlier than the learning processes ordinarily described as the parent's major contribution to a child's development. They involve processes that are distinct from learning and at the same time go considerably beyond the mere provision of sufficient nutrient and warmth to permit maturation. They include some of the processes underlying Harlow's classic demonstration (1958) of the vital role played by social interactions in the development of "affectional" systems, but they go beyond a single behavior category and specify a number of regulatory processes that affect physiological as well as behavioral development.

Apparently nature, in the form of evolutionary processes, has seen fit to utilize the early parent–infant interaction as a powerful regulator of early development. Mounting evidence that the basic regulators of development are not all within the infant organism is consistent with the intuitive belief that the early environment of the infant is somehow most critical for determining outcome. The survival value of this evolutionary development is likely to be the additional adaptability that is possible with this opportunity for cross-generational transfer. Events affecting parents prior to conception—for example, in response to a changing ecology—are theoretically capable of affecting basic physiological and behavioral traits of the young through modification of the parent's role in the early social relationship.

Finally, the existence of these early extrinsic regulatory processes may shed light on the biological function of the intense attachment so characteristic of infants and parents. The arrangement of having important regulators of development within the parent–infant interaction makes it imperative that the infant and parent be highly motivated to maintain a close relationship during this phase of development. The maladaptive consequences of lack of this early developmental regulation may have been an important selection pressure leading to the evolution of attachment and separation distress, in addition to the protection from predators conveyed by social bonds at all ages.

VIII. REFERENCES

Ackerman, S. H., Hofer, M. A., and Weiner, H. (1975). Age at maternal separation and gastric erosion susceptibility in the rat. *Psychosom. Med.* **37**:180–184.

Ader, R. (1969). Early experiences accelerate maturation of the 24 hour adrenocortical rhythm. *Science* **163**:1225–1226.

Ader, R., and Deitchman, R. (1970). Effects of prenatal maternal handling on the maturation of rhythmic processes. *J. Comp. Physiol. Psychol.* **71**:492–496.

Astic, L., and Jouvet-Mounier, D. (1968). Effets du sevrage en fonction de l'âge sur le cycle veille-sommeil chez le cobaye. *J. Physiol.* (Paris) **60**:389.

Bowlby, J. (1969, 1973). *Attachment and Loss, Vol. 1, Attachment* (1969), *Vol. 2, Separation* (1973), Basic Books, New York.

Brackbill, Y. (1971). Effects of continuous stimulation on arousal levels in infants. *Child Dev.* **42**:17–26.

Brazelton, T. B., Tronick, E., Adamson, L., Als, H., and Wise, S. (1975). Early mother–infant reciprocity. In *Parent–Infant Interaction*, CIBA Foundation Symposium 33, Elsevier, New York, pp. 137–149.

Bronson, F. H. (1974). Pheromonal influences on reproductive activities in rodents. In Birch, M. C. (ed.), *Pheromones*, Elsevier, New York, pp. 344–365.

Campbell, B. A., and Mabry, P. D. (1973). The role of catecholamines in behavioral arousal during ontogenesis. *Psychopharmacologia* **31**:253–264.

Campbell, B. A., Lytle, L. D., and Fibinger, H. C. (1969). Ontogeny of adrenergic arousal and cholinergic inhibitory mechanisms in the rat. *Science* **166**:637–638.

Carlier, E., Nowaczyk, T., Valatx, J. L., and Juvanez, P. A. (1974). Étude du sommeil du raton nouveau-ne isolé de sa mère; effets de l'alpha-methyl-dopa. *Psychopharmacologia* **37**:205–215.

Condon, W. S., and Sander, L. W. (1974). Neonate movement is synchronized with adult speech: Interactional participation and language acquisition. *Science* **183**:99–101.

Fairfield, J. (1948). Effects of cold on infant rats: Body temperatures, oxygen consumption, electrocardiograms. *Am. J. Physiol.* **155**:355–365.

Feldman, M. W., and Lewontin, R. C. (1975). The heritability hang up. *Science* **190**:1163–1166.

Flandera, V., and Novakova, V. (1974). Effect of mother on the development of aggressive behavior in rats. *Dev. Psychobiol.* **8**:49–54.

Fleming, A. S., and Rosenblatt, J. S. (1974). Olfactory regulation of maternal behavior in rats. II: Effects of peripherally induced anosmia and lesions of the lateral olfactory tract in pup-induced virgins. *J. Comp. Physiol. Psychol.* **86**:233–246.

Gaensbauer, T. J., and Emde, R. N. (1973). Wakefulness and feeding in human newborns. *Arch. Gen. Psychiat.* **28**:894–897.
Hall, W. G., Cramer, C. P., and Blass, E. M. (1975). Developmental changes in suckling of rat pups. *Nature* **258**:318–320.
Harlow, H. F. (1958). The nature of love. *Am. Psychol.* **12**:673–685.
Hinde, R. A. (1975). Mothers' and infants' roles: Distinguishing the questions to be asked. In *Parent-Infant Interaction,* CIBA Foundation Symposium 33, Elsevier, New York, pp. 5–16.
Hofer, M. A. (1970). Physiological response of infant rats to separation from their mothers. *Science* **168**:871–873.
Hofer, M. A. (1973a). The effects of brief maternal separations on behavior and heart rate of two week old rat pups. *Physiol. Behav.* **10**:423–427.
Hofer, M. A. (1973b). Maternal separation affects infant rats' behavior. *Behav. Biol.* **9**:629–633.
Hofer, M. A. (1973c). The role of nutrition in the physiological and behavioral effects of early maternal separation on infant rats. *Psychosom. Med.* **35**:350–359.
Hofer, M. A. (1974). The role of early experience in the development of autonomic regulation. In DiCara, L. (ed.), *The Limbic and Autonomic Nervous System: Advances in Research,* Chap. 6, Plenum, New York, pp. 195–221.
Hofer, M. A. (1975a). Studies on how early maternal separation produces behavioral change in young rats. *Psychosom. Med.* **37**:245–264.
Hofer, M. A. (1975b). Survival and recovery of physiologic functions after early maternal separation in rats. *Physiol. Behav.* **15**:475–480, 1975.
Hofer, M. A. (1976). The organization of sleep and wakefulness after maternal separation in young rats. *Dev. Psychobiol.* **9**:189–206.
Hofer, M. A., and Reiser, M. F. (1969). The development of cardiac rate regulation in preweanling rats. *Psychosom. Med.* **31**:372–388.
Hofer, M. A., and Weiner, H. (1971). The development and mechanisms of cardiorespiratory responses to maternal deprivation in rat pups. *Psychosom. Med.* **33**:353–362.
Hofer, M. A., and Weiner, H. (1975). Physiological mechanisms for cardiac control by nutritional intake after early maternal separation in the young rat. *Psychosom. Med.* **37**:8–24.
Hofer, M. A., Shair, H., and Singh, P. (1976). Evidence that maternal ventral skin substances promote suckling in infant rats. *Physiol. and Behav.* **17**:131–136.
Hoffman, H. S., and Solomon, R. L. (1974). An opponent-process theory of motivation. III: Some affective dynamics in imprinting. *Learn. Motiv.* **5**:149–164.
Klopfer, P. H., and Gamble, J. (1966). Maternal "imprinting" in goats: The role of chemical senses. *Z. Tierpsychol.* **23**:588–592.
Korner, A. F., and Thoman, E. B. (1970). Visual alertness in neonates as evoked by maternal care. *J. Exp. Child Psychol.* **10**:67–68.
Kravitz, H., and Boehm, J. J. (1971). The effects of institutionalization on development of stereotyped and social behaviors in mental defectives. *Child Dev.* **42**:399–413.
Kuffler, S., and Nichols, J. G. (1976). *From Neuron to Brain,* Sinauer Assoc., Sunderland, Mass.
Leifer, A., Leiderman, P. H., Barnett, C., and Williams, J. (1972). Effects of mother-infant separation on maternal attachment behavior. *Child Dev.* **43**:1203–1218.
Leon, M. (1974). Maternal pheromone (Monograph). *Physiol. Behav.* **13**:441–453.
Leon, M. (1975). Dietary control of maternal pheromone in the lactating rat. *Physiol. Behav.* **14**:311–319.
Leon, M., and Moltz, H. (1972). The development of the pheromonal bond in the albino rat. *Physiol. Behav.* **8**:683–686.

Levin, R., and Stern, J. M. (1975). Maternal influences on ontogeny of suckling and feeding rhythms in the rat. *J. Comp. Physiol. Psychol.* **89**:711–721.

MacFarlane, J. A. (1975). Olfaction in the development of social preferences in the human neonate. In *Parent–Infant Interaction*, CIBA Foundation Symposium 33, Elsevier, New York, pp. 103–113.

Mason, W. A., and Berkson, G. (1974). Effects of maternal mobility on the development of rocking and other behaviors in rhesus monkeys: A study with artificial mothers. *Dev. Psychobiol.* **8**:197–211.

Plaut, M. (1970). Studies of undernutrition in the young rat: Methodological considerations. *Dev. Psychobiol.* **3**:157–167.

Reite, M., Kaufman, I. C., Pauley, J. D., and Stynes, A. J. (1974). Depression in infant monkeys: Physiological correlates. *Psychosom. Med.* **36**:363–367.

Roffwarg, H. P., Muzio, J. N., and Dement, W. (1966). Ontogenic development of the human sleep–dream cycle. *Science* **152**:604–619.

Rosenblatt, J. S. (1971). Suckling and home orientation in the kitten: A comparative developmental study. In Tobach, E. Aronson, L. and Shaw, E. (eds.), *The Biopsychology of Development*, Academic Press, New York, pp. 345–410.

Rosenblatt, J. S. (1975). Prepartum and postpartum regulation of maternal behavior in the rat. In *Parent–Infant Interaction*, CIBA Foundation Symposium 33, Elsevier, New York, pp. 17–31.

Sander, L. W., Julia, H. L., Stechler, G., and Burns, P. (1972). Continuous 24 hour interactional monitoring of infants reared in two caretaking environments. *Psychosom. Med.* **34**:270–282.

Schleidt, W. M. (1973). Tonic communication: Continual effects of discrete signs in animal communication systems. *J. Theor. Biol.* **42**:359–386.

Scott, J. P. (1962). Critical periods in behavioral development. *Science* **138**:949–958.

Seegal, R. F., and Denenberg, U. H. (1974). Maternal experience prevents pup killing in mice induced by peripheral anosmia. *Physiol. Behav.* **13**:339–341.

Sharma, K. N., and Nasset, E. S. (1962). Electrical activity in mesenteric nerves after perfusion of gut lumen. *Am. J. Physiol.* **202**:725–730.

Singh, P. J., and Tobach, E. (1975). Olfactory bulbectomy and nursing behavior in rat pups (Wistar DAB). *Dev. Psychobiol.* **8**:151–164.

Singh, P. J., Tucker, A. M., and Hofer, M. A. (1976). Effects of nasal $ZnSO_4$ irrigation and olfactory bulbectomy on rat pups. *Physiol. Behav.* **17**:373–382.

Solomon, R. L., and Corbit, J. D. (1974). An opponent-process theory of motivation. I: Temporal dynamics of affect. *Psychol. Rev.* **81**:119–146.

Spitz, R. A. (1945). Hospitalism: An enquiry into psychiatric conditions in early childhood. *Psychoanal. Study Child* **1**:53–80.

Stern, D. A. (1974). Mother and infant at play: The dyadic interaction involving facial, vocal and gaze behavior. In Lewis, M. and Rosenblum, L. (eds.), *Origins of Behavior: The Effect of the Infant on Its Caregiver*, Vol. 1, Wiley, New York.

Stone, E., Bonnet, K., and Hofer, M. A. (1976). Survival and development of maternally deprived rats: Role of body temperature. *Psychosom. Med.* **38**:242–249.

Teicher, M. H., and Blass, E. M. (1976). Suckling in neonatal rats: Eliminated by nipple lavage; reinstated by pup saliva. *Science* **193**:422–424.

Terkel, J., and Rosenblatt, J. S. (1971). Aspects of non-hormonal behavior in the rat. *Horm. Behav.* **2**:161–171.

Thoman, E. B., and Korner, A. F. (1971). Effects of vestibular stimulation on the behavior and development of infant rats. *Dev. Psychol.* **5**:92–98.

Vandenberg, J. (1969). Male odor accelerates female sexual maturation in mice. *Endocrinology* **84**:658–660.

Weiss, J. M., Glazer, H. I., Pottorecky, L. A., Brick, J., and Miller, N. E. (1975). Effects of chronic exposure to stressors on avoidance–escape behavior and on brain norepinephrine. *Psychosom. Med.* **37**:522–534.

Wenzel, B. M., and Zeigler, H. P. (eds.). (1977). *Tonic Functions of Sensory Systems*, Annals of the New York Academy of Sciences, Vol. 290, New York.

Chapter 8

SOCIAL BEHAVIOR ON ISLANDS

Robert A. Wallace

Department of Zoology
Duke University
Durham, North Carolina 27706

I. INTRODUCTION

Islands have been called natural biological laboratories—places where one can find numerous experiments in progress at any time. These experiments have rarely been of human design but that is of no matter. We are free to make certain assumptions about the nature of the experiment and then to monitor the results. The results have, in sum, provided us with a fortunate view of the mechanisms of natural selection. It should be noted that we really don't know much about islands and how they vary from the mainlands and from each other, but we do have a certain working knowledge of certain aspects of their ecology. This information is now beginning to furnish us with a certain insight regarding the influence of environmental factors on behavior, but perhaps the most neglected of these considerations is how the peculiar nature of the island environment can affect social behavior. It is part of the larger question of how the environment can mold social systems, producing parallel evolution and thus attesting to the pervasiveness and intensity of the environmental influence. The goal of this paper is to focus attention on this neglect and to illustrate a few of the ways that the study can begin to proceed. It is first necessary to establish the nature of the island niche.

Islands may be classified as belonging to one of three types, each with its own broad influences on its inhabitants. Continental islands are those that were disassociated from the mainland during the Pleistocene by a rise in sea level as the great glaciers melted about 1.2 million years ago, and they include the islands of Britain, Ireland, Borneo, Sumatra, Java, and the Falklands. Oceanic islands harbor fewer species (Lack, 1976) but were born more dramatically by rising volcanically from the seabed, such as did the Aleutians, the Galápagos, and the Hawaiian archipelago. This group also includes the fringing archipelagos, such as the Philippines, the western Pacific islands, and, of particular interest here, the West Indies. These lie close to continents and receive the fringes of continental animal species. Most of these islands appeared in the Tertiary, but we have journalistically (as opposed to historically) recent examples as well. And finally there are the great islands. These are very old and large and were separated from the main continents by continental drift. Examples are Madagascar and Australia. However, since, in effect, all continents are surrounded by water, the island designation is essentially arbitrary, and so I will disregard the latter group. I will concentrate primarily on the oceanic islands since they are often very different from the nearest continents, and after all, the value of islands to evolutionary studies is related to the degree to which they are novel.

II. BEHAVING IN THE CARIBBEAN

Many of the examples here are drawn from West Indian species, so perhaps a brief review of the ecology of these islands is in order. The Greater Antilles are composed of Cuba, Jamaica, Puerto Rico, and Hispaniola, and the Lesser Antilles are represented by a strand of younger, smaller islands arching southward to Trinidad. The islands have been the sites of a recent series of investigations on the biology of certain species, particularly birds and lizards.

The larger Caribbean islands are generally divided into three altitudinal zones: the arid lowlands, the wet midlands, and the montane regions. The first two zones harbor the greatest number of resident species, with more species tending to breed in the arid lowlands. Speciation has been much more rapid in the more recently formed highland forest, as it is invaded by populations from the lowlands, indicating that selection is stronger there. (It is interesting that on the tropical mainland, there are also generally more species in the humid lowlands than in the montane forests, the difference being somewhat less on islands of comparable altitude.) Whereas the lowlands and midlands of islands do not differ greatly from the terrain at those altitudes on the mainland, the island highland forests are quite different from continental ones (Lack 1976). As an aside, there are fewer brightly colored insect species at higher elevations. It may be that there is selection for darker, heat-absorbing colors at these altitudes. One might wonder if the reduced conspicuousness at higher altitudes has had an impact on visual signaling in these species.

Lack (1976) theorized regarding the sequence of the colonization of Antillean islands by birds. He assumed that as Jamaica rose from the sea, coastal and marsh birds settled first; later came birds of the arid lowland shrubs and, as the island continued to rise, birds of the highland forest. In spite of the greater age of the lowland areas, there are no endemic resident water or marsh birds, but endemics do exist in the highland.

An understanding of the physical features of islands is important to an understanding of the ecological effects on social systems in such places. For example, because of the more intense selection in the highland areas, it is here that we should probably look for shifting social behavior that has been mediated by genetic change. The lowland areas, on the other hand, provide the best places to see the social effect of environmental influences as they select for behavior from within the preexisting behavioral spectrum. The courtship behavior of colonizing seabirds, as a case in point, probably has not changed much because their dispersibility has enabled them to interbreed freely with other groups. However, since their island nests are not likely to be exposed to predation, they may tend to nest closer together. Any such congregating would probably have a number of secondary effects, such as influencing parent–young relationships, since there would be a greater opportunity for the young to mingle.

Of all the Antillean islands, Hispaniola is the largest and highest. It lies 650 km northeast of the nearest mainland, Honduras. It was never connected to Honduras or to the other islands, but about 10,000 years ago it was only 400 km from Jamaica—as opposed to 700 km today—with three large, low islands between. At that time, Central America was temperate, with the bird fauna probably very similar to that found in the United States today (Lack, 1976). Honduras is thought to be the mainland source area for many of the West Indian species; however, most Antillean land mammals (comprised primarily of rodents, two opossums, armadillos, agoutis, and raccoons—with the introduced mongoose) came from North America (Gunderson, 1976). It has been suggested that the success of the mongoose, introduced in the late 19th century, illustrates that ecological gaps exist on islands, particularly for mammals. The gaps, in the case of mammals, may well exist because these animals are such poor travelers. It seems likely that poor dispersers on islands are likely to encounter less competition than good travelers and hence to show magnified shifts in their ecology and social patterns.

In summary, then, islands confront colonizers with different sorts of environments from the ones they left on the mainland. The lowland and midlevel areas are not likely to be drastically different (with the exception of the lack of extensive riparian habitats on islands—a possibly significant difference [Janzen, personal communication, 1976]), but the montane areas of islands have less in common with the higher mainland elevations. The most successful island colonizers are easily dispersed generalists, which usually emigrate from marginal habitats on the mainland. Island colonists usually find few competing species and few predators, especially terrestrial predators.

III. ISLAND INFLUENCES ON BIOTA

In their benchmark treatises on islands, MacArthur and Wilson (1963, 1967) pointed out that islands have fewer species than continents, both plant (see Lack, 1970) and animal. There are a number of ways islands can reduce the kinds of their inhabitants. For example, Klopfer (1959) suggested that the number of taxa in any geographical area is dependent on the time elapsed since its colonization, its topographic variability, its geographic extent, and its climatic variability. For islands, the first three of these, and to a lesser extent the fourth, are associated with the distance from the mainland. Islands closer to continents (near islands) are more likely to be larger, higher, and more heterogeneous and to harbor a greater number of species than remote islands. The latter are usually small and low-lying, with few adaptive zones, and inhabited by few species. Of course, *remote* is a relative term (Matthews and Matthews, 1970). Birds may find many islands quite accessible, islands that would be virtually unreachable by large mammals. Lack (1976) pointed out that it is not really known whether more remote islands are ecologically poorer than nearer islands of similar size, but since they have fewer species of land organisms and more restricted habitats, they seem to be. MacArthur and Wilson (1967) noted gaps in the species of island birds when the island is remote from others. (Diamond, 1972, noted the same phenomenon for insular mountain species.) The gaps were attributed to random extinction.

The importance of island size was reiterated by Soulé (1972) and Soulé and Yang (1973), who found that the log of island area is the best predictor of the number of lizard species (and their morphological variation—an important point) within local populations on islands. Terborgh (1973) found that the number of bird families on islands diminishes with the island area. Some investigators, however, have found that indices of environmental diversity, such as the numbers of plant species in the habitat, account for the number of bird species better than does island area (Watson, 1964; Power, 1972). So island size may be important in another way. It has been correlated with elevation and the number of perennial plant species (Case, 1975). In general, then, larger, higher islands tend to harbor more species, partly because higher areas are wetter (Carlquist, 1965). Distance, then, may not be as important in the limitation of species as was once believed. In spite of the fact that plants probably do not have difficulty in reaching remote islands, these places may have few plant species because they are almost always small and of uniform climate (Lack, 1976). It has also been pointed out that the reduced number of plant species on islands is correlated with fewer bird species (Lack, 1970). (For a general review of island biogeography see Darwin, 1859; Wallace, 1895; Carlquist, 1965; MacArthur and Wilson, 1967; Lack, 1976.)

At this point in our island investigations, we have no reason to reject the ecological operating principle correlating latitude with species diversity. Thus, all other things being equal, we can expect more species on tropical than on temperate islands.

I have mentioned the numbers of species on islands for two reasons: first, because they serve as a rough indicator of the suitability of an island for colonization by mainland forms, and second, because I will stress that the interspecific milieu is a strong determinant of social systems of the species comprising it.

In general, then, islands are relatively biologically depauperate places. Lack (1976) deliberately kept his definition of ecological poverty vague, but he included in his assessment factors such as habitat type, climate, vegetation type, altitudinal variation, the number of layers of canopy trees and bushes, and the food supply (size, edibility, and seasonal abundance of insects, fruits, and seeds). Hopefully, we will at some future date be able to describe these factors with more confidence, but for now we much rely on scattered data, impressions, and hunches, while generally assessing ecological richness by counting the organisms living in a circumscribed part of any island and not defining these areas too dogmatically. Unfortunately, much of the lacking data are precisely what we need to draw behavioral correlates of habitat and social behavior and, eventually, to ascribe causal relationships within the organism–environment complex. So at this point, we must phrase our findings in tentative terms and offer only qualified predictions of what behavior we can expect to find on any island type. The recent difficulty in correlating primate social systems to habitat type emphasizes the need for wide representative analysis in such attempts. Certain kinds of animals, however, more readily lend themselves to such problems. There is reason to believe that habitat influences are more quickly and perhaps more clearly reflected in the behavior of more "stereotyped" animals, such as lizards and birds, so it is primarily from among these groups that I draw most of the examples here.

But why should the behavior of island colonizers be of special significance to the study of social behavior? Islands are important zoogeographically because many of the biological principles operating in such places apply more or less to all natural habitats. Islands, however, are smaller, newer, and simpler than the vast continents and are biologically, as well as physically, more unstable. Thus the systems are, theoretically, easier to describe while being more sensitive and reactive. MacArthur and Wilson (1967) stated that the fundamental processes of dispersal, invasion, competition, adaptation, and extinction apply to island species as well as to those of continents but are easier to observe on islands. (They might have included processes influencing social behavior as well.) We find ourselves currently coping with the refinement of the principles of island ecology, and when we are not restricting ourselves to correlating physical variables, our attention is often drawn to the genetics of island species. Because of these investigative biases we have developed theories of island biogeography that account for the numbers and kinds of colonizing species, and although many of these theories are speculative (Lack, 1976), they do have some predictive value (Simberloff, 1976). With this groundwork, however uneven, we are now in a position to broaden our investigation to include one more variable: the changing social patterns of island colonizers and the environmental factors that drive those changes.

The social behavior of any species can be expected to be strongly influenced by its interspecific competitive milieu, and as I mentioned, the competitive picture is simplified in a species-poor environment. The paucity of species on islands has been attributed to two factors. The first assumes that the main problem in colonizing an island is reaching it. MacArthur and Wilson (1967) argued that land animals reach islands rarely or only by accident. The number of successful colonists is balanced by species that disappear through random extinction, a process that is particularly prevalent on small islands. (For reviews see Serventy, 1951; Preston, 1962; Mayr, 1963; MacArthur and Wilson, 1963; 1967; Hamilton, 1964; Selander, 1966; Lack, 1976.)

This explanation was challenged by David Lack (1976), who argued that the ecological poverty on islands is due to successful colonizers' excluding competing species. Lack assumed that islands are probably continually barraged with continental species but that most are repelled by the "broad shields" of earlier, and competing, colonists. Lack's hypothesis furnishes more material for the development of special theories to account for the social behavior of animals on islands. Lack assumed that a small number of generalized species tend to exclude through competition a greater number of more specialized species. Generality in behavior may be a critical feature for colonists, as will be pointed out.

If, as Lack (1970) argued, dispersal to islands by land birds is so frequent, why are there so many endemics? Lack attributed the high endemism to the strong selection on islands. He assumed that islands must be very different from continents, rendering selection so strong that it counteracts the potential genetic mixing associated with incoming birds from the mainland. In addition, he stated, local adaptations can take place much faster than previously believed. As an example of the intense selective pressure on islands, he pointed out that the islands of Fernando Po and Zanzibar are located the same distance from tropical Africa but that on the former, 30% of the bird species are endemics and on the latter, 3%. The difference, Lack believed, lies in the fact that Fernando Po is ecologically more different from the mainland than is Zanzibar.

Whereas it seems that Lack's arguments may be essentially correct, the influence of dispersibility on the competitive environment of any colonizing species cannot be minimized. Because of differences in the ease with which various kinds of animals travel to islands, the resulting social situation may be very different for each kind. Birds travel easily and island colonizers can expect to have competitive problems with incoming species of similar adaptive type. True freshwater fish rarely make it to islands; neither do heavy seeds of trees, while reptiles and mammals have even more trouble. If a species belonging to any of these groups should manage to establish itself on an island, it may have far less ecological and reproductive interference from competitors.

IV. BEHAVIORAL TRAITS OF COLONIZERS

In a sense, then, islands select their colonizers in that they can be reached more easily by some kinds of species than by others. But once they are reached, what kind of organism is more likely to establish a toehold? Obviously, an animal that finds a habitat similar to the one it left will be more successful, but that likelihood becomes more remote as the island does. So, to some degree, a hopeful colonizer is likely to encounter novel conditions. Therefore, it is more likely to succeed if it is behaviorally flexible and/or a generalist. Flexible generalists usually stem from unstable or peripheral populations, not from the center of large, complex, and well-adapted mainland populations (Darlington, 1970). This argument is supported for insects by Wilson (1959, 1961), who found that among ants, the best colonizers occupy marginal habitats on the mainland (i.e., open lowland forest, savanna, monsoon forest, and littoral zone). Such species also tend to occupy a broader range of habitats on the continent. Later, Janzen and Schoener (1968) found that beetle samples from islands off Central

America were not a random sample from the continental lowlands. Whereas predatory insects constituted only 3–15% of the species on the mainland, ladybird beetles (Coccinellidae), which are predatory generalists, alone made up about 40% of the island insect species. Spiders and ants were also common on the islands, and they, too, have a great ability to exploit sporadic and fluctuating food of a variety of types (Janzen, 1973). Williams (1969) concluded from his long studies of West Indian lizards that the most successful colonizer, *Anolis carolinensis,* is not a stable, deep-forest species but is tolerant of a variety of conditions that individuals from the more stable species cannot survive. Case (1975) described the best lizard colonizers of islands in the Gulf of California as capable of reaching high densities, having high birth and death rates, and being habitat generalists. Among birds, second-growth species are good colonizers and persisters (as opposed to deep-forest species) apparently because they are adapted to changing habitats on the mainland, as their second-growth habitat disappears at one place, being replaced by forest, and then reemerges at another place, where some field has been abandoned. So, among birds, second-growth species are good island colonizers because they must frequently move and must be able to tolerate at least a modest variety of habitats (MacArthur, 1972). We find, then, that good island colonizers tend to be generalists and opportunists whose flexibility apparently stems from their having been previously adapted to marginal habitats on the mainland.

It is important to understand, however, that if a species is ecologically tolerant, its social behavior is not necessarily correspondingly unspecific. In fact, the two behavioral traits can evolve quite independently (see Wallace, 1973). I hope to illustrate that broad ecological tolerance has its effect on social behavior primarily by increasing the range of adaptive zones in which a population exists, as each zone then, in turn, exerts its own particular pressures, thereby increasing variation in the social systems of its denizens.

The simultaneous selection for ecological generalists and mating "specifists" is particularly adaptive if islands are emphatically special places. Since, by definition, islands are reachable if they are inhabited, it may be to a colonizing population's advantage to shift quickly to a social system that would exclude new conspecific colonists from joining the breeding unit and thereby diluting the concentration of the gene pool around the island optimum. In other words, continued interbreeding with incoming representatives of the parent population can retard directional selection. Thus, the individuals in the colonizing group, once the group is large enough to maintain itself, might quickly become reproductively more specific while retaining or increasing their ecological flexibility. The necessity for quickly developing a restricted breeding system on an island would depend upon whether the island was rarely reached by the parent population or was continually barraged by conspecific propagules, the necessity being stronger in the latter case.

V. THE EFFECT OF WIDE NICHES ON BEHAVIOR

I have described island colonizers as encountering few competitors and have mentioned two explanations advanced to account for the phenomenon: MacArthur and Wilson's (1967), which stressed the problems in reaching islands, and Lack's (1976), which assumed that propagules were usually outcompeted by previously established colonizers of similar adaptive types. The latter argument is supported by the observation that generalists usually make the best colonizers since, by being able to opportunistically diverse into a number of habitat types or ecological roles, a population of generalists will be able to exploit enough resources to maintain itself, while, by occupying a wide range of "niches" and becoming better adapted to them, it excludes incoming propagules from being able to occupy those niches. It might be asked: How could islands, with so few species, provide the diversity necessary for niche expansion? However, increased opportunity for niche diversity based on complexity (structural diversity) of a resource need not be entirely dependent upon a high number of species' comprising that resource, a finding that has been substantiated for birds (Orians, 1969) and vegetation (MacArthur, 1969).

It has been shown, in support of the wide-island-niche argument, that many species of island birds are found in more kinds of habitats than the same species on the mainland (Lack and Southern, 1949; Marler and Boatman, 1951; MacArthur and Wilson, 1967). (For exceptions, see Crowell, 1962; Pulliam, 1970; Soulé and Yang, 1973.) Dispersion through a range of habitats is only one measure of niche width, however, as I shall point out. As further evidence of wider niches on islands, Grant (1969) suggested that West Indian bird species are not derived from a random sample of mainland species but from those on the mainland that differ markedly in bill size. Among birds, colonizers apparently exclude subsequent propagules of similar bill size. Lack (1976) generally attributed such disjunctions in size of trophic structures of island species to simple character displacement, but as will be pointed out, there are a number of social pressures that can account for such shifts.

Birds offer other indications that island niches are broader. First, there are a number of cases in which intermediate-sized bird species replace large and small species on islands and then expand their niche to fill both ends of the ecological spectrum (Lack, 1976). Second, because of their mobility, birds are able to migrate, many species pouring into southern climes each winter, but more migrate to continents than to islands, perhaps because it is easier to make room among many specialists than among a few generalists. The greater the number of species in an area, the more ecological interfaces there will be, each interface populated by individuals that are probably not well adapted to their group's ecological periphery. Consider the parulid warblers. Of the 21 species of birds that appear on Jamaica each fall, 18 are parulid warblers. Of these, 17 (the other being scarce) segregate ecologically (throughout differences in habitat, feeding stations, and foraging mode—indicating that the ecological shield against new propagules is not totipotent), but 12 of these are transients, leaving only 5 to winter there. However, 30 parulids winter in Honduras, apparently wedging themselves between the resident specialists, further, if temporarily, subdividing the habitat (Lack, 1976). It may be argued that southbound migrants prefer continental areas because they have historically traveled to such places, perhaps before the islands were born.

Any change in the numbers of interspecific competitors undoubtedly has pronounced effects on the social pressures on the competing species, but these have not yet been extensively investigated. Birds, because of seasonal changes in species abundance, present an unusual opportunity to monitor the effects of such changes, but so far only broad shifts have been described, such as the formation of mixed winter flocks. (The major problem, of course, lies in controlling or accounting for seasonal changes other than shifts in species number.) In any case, interspecific competitors undoubtedly cause changes in social systems. For example, as interspecific competitors increased, a species might be forced into a narrower niche where the sexes would be encouraged to interact more frequently and perhaps to subdivide the niche. Increasing competition might also necessitate interspecific territoriality. Interspecific competition could be expected to alter (reduce) the numbers of individuals in a population and thus to have a pronounced effect on the social behavior of the group. Any such social effects of high numbers of interspecific competitors would, of course, be attenuated on islands.

It must not be assumed, however, that competition is relaxed on islands, at least for established species. Any relaxation is of a temporary nature and can last only until the population increases to the point where it encounters barriers, food or otherwise. It may rather quickly encounter shortages in some commodity on an island if the island has been colonized by few food species. Species that are not closely preadapted to the island condition may fall into strong competition earlier in the colonizing period.

The response for any colonizer may be to shift into new, unoccupied adaptive zones, but these, too, may become quickly "saturated" because of the paucity of other species that that colonizer might exploit. Thus, whereas competitors may be few, competition may be high, as I mention in the following discussion of shallow island niches.

VI. AGGRESSION ON ISLANDS

It has been argued that colonizing species are "ecologically aggressive" (see Wallace, 1973), but little has been said about the behavioral aggression of island colonizers. However, it appears that many of these species are, in fact, more aggressive toward each other than are their mainland counterparts. For example, Pulliam (1973) found decreased sociality and higher aggression in the Fringillidae, Thraupidae, and Icteridae of Jamaica. Of these families in Jamaica, 18% showed some social tolerance (such as flocking or the formation of family groups) compared to 68% in Costa Rica. He also found mixed species flocks common in all habitats in Costa Rica but totally absent on Jamaica. In addition, the Jamaican grassquits were found to be less social and more territorial than their Costa Rican relatives.

It may be that increased aggressiveness is a mark of poorly preadapted recent colonizers or older, established populations. Poorly preadapted recent colonizers may be hard put, at first, to find commodities in sufficient quantities to maintain themselves. After becoming better adapted to their island homes, however, intraspecific territoriality may be less important, and other means, such as niche subdivision or a change in food size, may be employed to reduce aggressiveness based on competition and to ensure sufficient commodities. Old, established island populations, on the other hand, may be well adapted to the island niche and may show strong intraspecific aggression based largely on competition for commodities that are efficiently exploitable.

Higher aggression may also be a spacing mechanism that has the same adaptive basis on islands that it does on continents and may have little to do with insularity *per se*. Pulliam (1973) accounted for the low density and high aggression of some Jamaican birds by noting the high incidence of a common foot mite that is quickly spread among congregating birds. The incidence or prevalence of the mite on the island has not been related to the insularity of the environment. This finding clearly illustrates that behavioral correlations with insularity are not enough to establish an island effect on behavior.

VII. THE BEHAVIORAL EFFECTS OF INCREASED NUMBERS ON ISLANDS

A phenomenon sometimes called the *Krebs effect*[1] may operate on certain island species as evidenced by the fact that their numbers soar over the levels of their mainland source populations. Lack (1971) found passerine species abundant and widespread on Puerto Rico and Jamaica. (In fact, he saw 87% of the species on these islands in two or three days.) The 18 bird species of the island of Puercos in the Pearl archipelago have combined densities as great as over 60 species in comparable mainland habitats (MacArthur, Diamond, and Karr, 1972). Certainly, such increased density can be expected to be associated with social changes since the probability of social interaction is increased. Such changes might take the form of stronger pair bonding, which could withstand continual interruption by potential sexual competitors, or they might be reflected in more focused social signaling, which would produce stronger and more efficient threat and appeasement gestures.

Perhaps the Krebs effect does not operate on islands, at least in the way it was originally described. Krebs *et al.* referred to "small" islands and enclosures, but when the MacArthurs (MacArthur, R. H. *et al.*, 1973) compared a large island, Cañas (500 hectares) with the smaller Puercos (70 hectares), they found that the densities were about equal. It seems that these denser populations could not be maintained without social changes involving lowered aggression. It also seems, then, that such adjustments could appear only over rather long periods of adaptation and hence are a property of older colonizing populations, an idea that remains to be tested.

Janzen (1973) offered some ecological explanations for changes in numbers of plants and insects on islands. When he compared Costa Rica to the islands of Puerto Rico, San Juan, St. John, Providencia, San Andrés, Greater St. James, Icacos, Steven Cay, and Palominitos, he found that the island plants typically had relatively broad niches, as evidenced by their appearance in both wetter and drier habitats. There were, accordingly, fewer species of plants and insects on the islands than on the mainland. He noted that the large number of introduced plants have probably had little impact on diversifying the system because of their recent arrival and that

---

[1] Krebs, Keller, and Tamarin (1969) enclosed two-acre fields with wire-mesh fences topped with sheet metal so that no mouse could enter or leave. Predation, however, was allowed to remain unchanged. The numbers of *Microtus* increased threefold, perhaps because they were not allowed to emigrate. These authors have suggested that normal population regulation cannot exist inside a small enclosure and that islands are, in effect, small enclosures.

they have generally not yet been adapted to by the local insects. Thus, herbivorous island insects are faced with a food supply of low diversity. However, whereas mainland insect populations generally swell in the rainy season, the island populations remain low in both the dry and the wet seasons (Janzen, personal communication, 1976). On the mainland, when species numbers are reduced by seasonal hardship, normally some of the most persistent species will burgeon, but this did not occur for beetles (Coleoptera) and bugs (Hemiptera), while it did occur for stem suckers (Homoptera). Janzen's explanation was that beetles and bugs are so specialized and inflexible that they do not easily respond to a seasonal lowering of competitors, whereas homopteran generalists, which feed on a wide variety of plants, would immediately respond to reduced competitive pressures.

The stable and low-density populations of plants and insects on islands provide two theoretical fulcra. First, these organisms occupy the lower levels of the terrestrial food chain, and their wide fluctuations on the mainland and continuous low densities on islands are undoubtedly responsible for many of the foraging and social differences in mainland and island vertebrates (Keast, 1970; Wallace, 1974b).

Second, the stable low numbers of these species on islands may indicate a shift toward K-selection on islands (MacArthur and Wilson, 1967; Terborgh, 1973). K-selection, the tendency to develop fewer, higher-quality offspring, develops in highly competitive situations, and the ecological poverty in which island colonizers find themselves can be expected to increase the level of competition for the limited commodities in some kinds of groups. A shift from r-selection (where numerous, lower-quality offspring are produced) to K-selection would have profound social implications. Under the latter conditions, organisms could be expected to put less energy into simple egg or gamete production (birds lay fewer eggs on islands) and more energy into caring for offspring. Such care could be expected to influence pair bonds (perhaps resulting in both parents' tending the young), territorial behavior (which might ensure a continued food supply for slower-developing, K-selected young), habitat expansion (in an effort to ensure a wider range of food types), and interspecific aggression (which might exclude food competitors).

Interestingly, K-selection on islands might be rapidly brought about by the arrival of r-selected propagules. As rapidly reproducing populations of colonizers swelled in response to r-selection, they would quickly encounter ecological barriers on the depauperate islands, such as the food barrier, bringing about a shift to the production of competitive, K-selected offspring.

VIII. THE RELATIVE SOCIAL EFFECTS OF COMPETITION
AND PREDATION ON ISLANDS

One might ask: Are those island populations that increase in number responding primarily to reduced competition or lower predation? It is assumed that the effects are not necessarily mutually exclusive, but this does not mean they cannot be analyzed independently, and, in fact, some studies support one argument to the exclusion of the other. Puercos Island was compared with mainland plots of similar structure, the Puercos forest being intermediate between the two mainland plots, moist forest and moist shrub. Puercos, with its 70 hectares, held only 20 of the mainland's bird species, 56 and 58 species found on each 2-hectare plot. (Larger areas would have held many more.) However, whereas the mainland plots held 0.33 and 0.28 pairs per species per hectare, the island held 1.35 pairs. In six cases where mainland and island species were compared, the Puercos birds were 5.6 times more common (Karr, 1971). It made no difference whether these species were protected hole-nesters, such as woodpeckers, or species that were more vulnerable to predators. MacArthur (1972) considered this information to be the first line of evidence that the increased density on islands is due to competitive, rather than predatory, release. He buttressed his argument by noting that the species that increased most was the barred antshrike, *Thamnophilus doliatus,* a kind of antbird that on the mainland had the greatest number of competitors.

Predators may not generally be a problem on islands. In particular, islands may present problems to large predators. Prey density lessens as the size of the prey species increases (Krebs, 1972), and on islands the general paucity of species magnifies the problems for any predator finding food within the larger food categories. An increase in abundance of certain species alleviates the problem to some extent, but fewer prey species, even if well-represented ones, mean a more unstable, and hence unreliable, food supply for any predator if they fall into seasonal short supply with no alternative prey available. In general, then, one may expect to find fewer large predators on islands, partly because they may have problems of dispersal (particularly mammalian predators) and partly because the peculiar composition of the food supply on islands may not be able to sustain them.

The social implications of a population increase on islands through release from predatory, as opposed to competitive, pressures are quite different. If the increase is due to release from competitors, species may become socially more flexible in that they might become more tolerant of other species, paving the way for, for example, interspecific feeding assemblages. If the other species offer no sexual competition, which would result in breeding mistakes, there would be an increased tolerance for eccentricity within the species, thus probably increasing the behavioral variation within the group.

On the other hand, if population increase is primarily due to a relaxation of predation pressures, a species would be free to become more conspicuous. Conspicuousness, of course, enhances communication as the animal and its message become less cryptic, and so increasingly complex signaling systems can develop that would permit higher levels of social coordination. Predator-free species can also be expected to be bolder (or tamer), and the social repercussions can be wide-ranging, for example, in adding behavioral boldness to the impact of visual conspicuousness.

IX. MULTIPLE INVASIONS, POPULATIONS, AND NICHE SIZE

It has been suggested that island colonizers tend to exclude incoming competitors, but they are not always successful. Mayr (1942) described "double invasions," the colonizing of an island by related stock from the same source. He suggested that they may be able to coexist but that the price is likely to be a narrowing of the niche for both species. Examples of such instances are cited in Moreau (1966).

There are other ways competing species can coexist on islands, but these usually involve a reduction in number for one of the species. The rarity exists whether one is considering numbers (Moreau and Ridpath, 1966; Parkes, 1965; Ripley and Bond, 1966; Watson, 1964) or standing-crop biomass (Grant, 1966c). If the rarity exists for a species occupying the lower trophic levels, the effect can be wide-ranging. Grant (1966b,c) reasoned that the smaller the total variety of food types on islands, the more likely they are to be unequally represented. This inequality, then, influences the biomass of other species that compete for that food, possibly resulting in an inequality in the biomasses of the competitors. If their biomasses should be the same, there will probably be numerical imbalance. If the imbalance is due to one species' becoming rarer, the rarity can be expected to produce social changes for the population in question. For example, among birds and lizards, sexual signaling might shift to longer-distance modes. Also, there might be less selection for intraspecific aggression coupled with decreasing tolerance for the numerically superior competitor species.

In the case of numerically imbalanced competitors, selection might minimize the imbalance by shifting the rarer species to a smaller size for energy advantages, particularly if the island were small (Grant, 1968). This change would partially enable the population density of the rarer species to recover. At the same time, changes could be expected in the sizes of territories since territory size is partly related to body size (Armstrong, 1965). As territories grew smaller, more social interaction could be expected since, within a circumscribed island, a higher density of territories would increase the number of individuals with which any animal shared its boundaries.

Two species of competitors can also coexist in any area if they tend to occupy scattered colonies (MacArthur, 1972). Individuals of each species would thereby interact more often with conspecifics than with interspecific competitors and would thus tend to inhibit themselves more than the competitor, through density-dependent mechanisms. If competing species were dispersed over an island they would exert powerful reciprocal effects, probably leading to the extinction of one of the species; however, they might avoid serious competition if they chose to live in strongly demarcated clumps.

The closer intraspecific association caused by clumping could be expected to have certain social advantages, such as promoting greater synchrony and coordination within the group. On the other hand, increased competitive interaction might also select for high intraspecific aggression. Intrasexual aggression could be expected where increased density continually threatened the disruption of pairs by the presence of potential sexual interlopers.

X. SHALLOW ISLAND NICHES

It seems apparent that niche broadening on islands can occur because of a lack of competitors. However, the impetus for moving into "unoccupied niches" is often more rapid than one might expect and the rapidity may be associated with the inadequacy of the island niche; in many cases, there may not be enough of a single *type* of commodity. Because of problems of colonization of food species, for example, a growing population may quickly deplete its food supply and be forced to seek new sources or look in new places, thus expanding its niche. The paucity of food species, food types (for example, foods within a size class), or food biomass for most island colonizers (Schoener, 1965) can be described as *niche shallowness* (Wallace, 1974b). On continents, on the other hand, problems of competition force species into narrow niches, but the abundance of food species ensures a greater food supply within each niche. The relationship of problems of colonization and niche depth can be conceptualized as in Figure 1. Niche shallowness, then, serves to explain why island colonizers might so

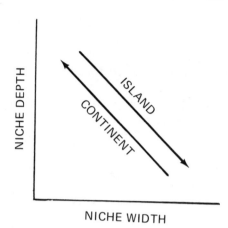

Fig. 1. The theoretical relationship between niche size on islands and continents. See the text for an explanation.

quickly fall into intraspecific competition of the sort that could force niche expansion.

XI. BEHAVIORAL CORRELATES OF NICHE BROADENING

It might be worthwhile to consider how niche broadening can occur on islands and what are the behavioral ramifications of such changes. Basically, niches can be broadened in five ways: through (A) range expansion; (B) habitat expansion; (C) increase in variation; (D) increase in sexual dimorphism; and (E) increase in size.

A. Range Expansion

The most obvious way a population can increase its niche width is to move into new geographical areas, such as onto an island, where it will undoubtedly be forced to interact with its environment in new ways. In the following examples, I refer specifically to niche changes on islands.

B. Habitat Expansion

A colonizing species will undoubtedly first seek out the island habitat most similar to the one it left. However, it may find no ecological counterpart of the mainland habitat on its island and be forced to live in new kinds of places. In addition, it may quickly be forced into new habitats because of strong competition in a shallow niche. Lack (1976) noted that habitat expansion may be documented by a cataloging of the number of island species that live in diverse kinds of places.

It might seem that island forms, forced to broaden their shallow niches partly as a result of strong competition, would tend toward exclusiveness in habitat occupancy. However, such is not always the case. In a comparison of mainland (Honduras) and island (Jamaica) bird species, Lack (1971) found that most congeneric species of passerines occupy separate habitats (115 of 222 species pairs, or 52%), whereas on the island only 2 (or 10%) of 20 possible congeneric pairs of species are separated by a difference in habitat. The range of habitats in the two areas are similar for only about one-fourth of the species.

The increased habitat sympatry on islands is partly a function of the simplicity of the habitat. For example, among birds, there is little separation by habitat in the Galápagos because there are only three distinctive types of woodland: the arid zone, the humid zone, and the locally isolated mangroves (Lack, 1971). Klopfer and MacArthur (1961) predicted that where there is high faunal diversity, animal species will have narrower niches with more niche overlap than will species in faunistically depauperate areas.

New habitats may be invaded in ways other than expansion of geographical area. For example, animals may shift to different altitudinal strata. In comparing Puercos with Costa Rica, MacArthur (1972) found that the island birds increased the range of their foraging height in the trees as they increased in abundance.

An increase in the range of habitats on a heterogeneous island could be expected to encourage adaptive radiation. It has been customary to view adaptive radiation in terms of species and races, but it may also occur within populations as the group expands into different subniches or adaptive subzones. Such changes have been primarily studied in *Drosophila*, in which chromosome polymorphism has been interpreted as an adaptation of a population to a heterogeneous environment (Dobzhansky, 1961, 1963, 1965), and in the recent flurry of isozyme electrophoresis studies (cf. Selander *et al.*, 1971). If the adaptive radiation is a response to selection for occupancy of different kinds of habitat, a certain degree of breeding specificity for each subpopulation might arise. The level of interbreeding between subpopulations may be far too high to permit speciation, but some preference for mates from within the group can be expected as each group adapts to its own particular habitat. Thus, once firm subpopulations were established, increased specificity in mate selection could be expected, especially for viscous populations.

When a population comes to occupy a number of habitats, some are likely to be "better" than others (e.g., have a higher carrying capacity). Thus, the population will be denser in some areas than others, and density, as has been pointed out, can affect social systems in specific ways. There is, as yet, no substantiating data to indicate social differences in subpopulations of groups of various densities.

Habitat expansion, it should be pointed out, does not necessarily imply niche expansion. Diamond (1970) investigated bird species on islands off New Guinea and found no niche shift in 23 of the 50 species he studied. However, 12 had expanded into new habitats, 13 had increased the range of their foraging altitudes by foraging on mountainsides, and 12 did not expand spatially but became more dense in the counterparts of their "preferred" mainland habitat. It is noteworthy that island-colonizing birds often expand from second growth into forests and, in general, into higher altitudes (which on islands are more likely to be forested). The social effects of invading forests, especially regarding changes in signaling mode associated with a move into a less visual habitat, have not been investigated.

In summary, if a species is not relegated to a particular habitat but comes to occupy multiple subzones, each subzone may exert its own molding effect on the phenotype, including the social behavior, of the population. We know something about the general social effects of living in certain kinds of habitats, particularly as they apply to communication modes. Less is known about the environment's effects on group structure, but the first steps have been taken in primate studies (cf. Crook and Gartlan, 1966).

C. Increase in Variation

According to the niche-width-variation model of Van Valen (1965), species in wide niches, such as exist on islands, should show increased variation. The implication is that the variation will be expressed both morphologically and behaviorally, but it should be made clear that these phenotypes are not necessarily well correlated. It is no simple matter to compare morphological variation with behavioral variation in a meaningful way. Morphological variation is an easily quantifiable expression of genetic-environmental interaction, but behavioral variation may exist as a potential—a true phenotype within either an individual or a population, but one whose full expression may never be demanded by the environment. The correlation between morphological and behavioral variability is so undependable, in fact, that one may not reflect the other at all. In the yellow-faced grassquit, *Tiaris olivacea,* mainland and island forms both show low variation, with no statistical difference, but the island birds are significantly more variable in their choice of seed size (Pulliam, 1973) (a finding not predicted by the "compression theorum" of MacArthur and Pianka, 1966, but in agreement with Van Valen's niche-width-variation model, 1965).

Good candidates, whether individuals or populations, for island colonization are likely to be malleable in behavior, malleability being associated with opportunism. Malleability, however, implies a wide range of behavioral potential, a keyboard along which environmental demands strike only certain, mainly harmonic, chords. Some notes are destined never to be played, and an observer may wrongly conclude that the keyboard is missing those notes. No matter what notes the environment may strike, however, a good colonizer is likely to be an organism with a wide behavioral range.

Once a population has successfully colonized an island, the slow process of changing the keyboard through evolution may begin. Such change can be expected to increase the overall behavioral variation of the population, especially for relatively viscous populations in heterogeneous habitats.

Again, we are led to the question of selective variation. Are certain phenotypes held within strict limits while others are left free to vary? On the Galápagos, the finch *Geospiza fortis* is sympatric with the smaller *G. fuliginosa* and the larger *G. magnirostris*. The intermediate *G. fortis* is highly variable in its bill size and overlaps with its competitors at both ends of its bill-size spectrum. It apparently is highly variable in its foraging behavior as well (Lack, 1971). We might well ask whether its social behavior is also more variable than that of its ecologically more restricted larger and smaller competitors. It is likely that the variation in bill size developed in response to ecological influences, such as variable food size, but if this ecological expansion brings *G. fortis* into closer contact with its congeners, increased specificity in mate selection strengthened by interspecific aggression based on competition may result.

On islands that lie very near their faunal-source mainlands, such as the Tres Marías (Grant, 1967), selection may be operating in an unusual way. In such cases, the frequency of propagules' reaching the island may be so high that any directional selection is swamped by incoming genes. The island colonizers may still show higher variation than their mainland stock, however, if the island simply "selects" the individuals with the broadest tolerance from among the newcomers, the result being that each new island generation is mostly comprised of first- or second-generation colonizers. In such cases, the colonizers would be expected to exhibit the same behavioral biases as did the mainland group insofar as the behavioral repertoire is insensitive to immediate environmental influences.

In some cases, island colonists may vary in a surprisingly wide range of characters. It seems that some populations might, as a response to competitive release, show high, and possibly unadaptive, variation in a number of areas. For example, the Puerto Rican woodpecker, *Melanerpes portoricensis,* demonstrates a remarkable variation in the attachment of the tongue apparatus (Wallace, 1974a). It may be that such variation is the result of the sudden release of competitive strictures, coupled with reduced canalization, which permits a relaxation in the control of a wide number of phenotypic characters, many perhaps going unnoticed. Aberrant or seemingly unadaptive behavioral variation among island species would probably appear only as statistical variation in behavioral analyses, but future investigators might well take note of the level of exceptional instances of behavior on islands and continents.

Social patterns may influence morphological variation in some cases. For example, there may be social strictures on bill-size variation among birds. Brown and Wilson (1956) pointed out that character displacement in the bill size of mainland birds may act as a reproductive isolating mechanism. Since birds strongly rely on features associated with the head in making discriminations at various levels within the order (see Wallace, 1973), bill size may be a species-recognition feature and hence would not be free to vary greatly. Lack (1947), for example, demonstrated that among certain Galápagos finches, a species can distinguish between stuffed specimens of its own and another species when the second species differs in bill size but is quite similar in plumage. Wilson (1965) noted that *G. fortis* showed little variation in bill depth and suggested that bill depth in this species may be a species-recognition feature (but see Grant, 1967).

D. Increase in Sexual Dimorphism

Increased sexual dimorphism has been described for a number of kinds of animal species on islands. The dimorphism is more pronounced for some characters than others. For example, it is particularly obvious in the trophic appendages of island birds, but the same group may show *reduced* sexual dimorphism in plumage, often due to reduced male garishness or the retention of immature plumage (Sibley, 1957; Grant, 1965b,c; Amadon, 1953; Mayr, 1942). The reduction in plumage differences for the sexes is accompanied by either an increase or a decrease in continuous variation in both plumage and song (Crook, 1961; Grant, 1966a; Thielcke, 1965). The two conflicting trends in variation on islands indicates that there is not a simple "insular effect" on variation but that variation for any character is determined by some factor secondarily associated with insularity, whether social or "ecological."

It should be noted that whereas many species of vertebrates are statistically sexually monomorphic, nevertheless the male is usually larger in absolute terms and may therefore have larger trophic structures on the basis of allometric growth differences. These differences may provide the basis for the subsequent direction of sexual divergence as they are amplified through selection. Even on the mainland, the small differences in size of the sexes could be important in permitting slightly different foraging at times of critically low food supply. We have no way of knowing this at present, however, because our methods of analysis are far too crude to reveal any but the broadest characteristics of foraging behavior. The larger size of continental males, then, may not appear statistically significant, but it may be ecologically significant if it provides a small, helpful, perhaps critical, margin for a seasonally besieged population. We are primarily interested, however, in why such small sexual differences so often come to be magnified on islands.

Large size in male vertebrates may be due to sexual selection. Large males may arise because females tend to choose those males that demonstrate the superiority of their non-sex-linked genes (Trivers, 1976). Maynard Smith (1971), Williams (1975), and Zahavi (1975) pointed out that female choice is a central factor in the development of larger males and male–male competition (but see Grant, 1968). Perhaps the male's large size places him at a disadvantage in his environmental interactions, but this may be the burden he bears in exchange for the female's developing and laying the eggs.

A number of bird and lizard species are sexually dimorphic on islands, especially in their trophic structures. Not surprisingly, the size of the trophic structures can be related to foraging behavior. For example, the Puerto Rican hummingbird, according to the Keplers, is sexually dimorphic, and the sexes forage in different ways (cited in Lack, 1976). The extinct huia of New Zealand was sexually dimorphic, and the sexes were known to forage differently (Lack, 1944). In the Lesser Antilles a trembler, on Hawaii three sicklebills, and on São Tomé a sunbird are all sexually dimorphic in bill size (Lack, 1976), but we lack specific information regarding their foraging behavior.

The specificity of trophic structures as targets for natural selection is demonstrated in those bird species in which the bill changes disproportionately more than general body size or other structures. The western grebe, *Aechmophorus occidentalis,* is sexually dimorphic in bill size and shape, the females having smaller, upturned bills. They are sexually monomorphic in head size and wing length, so apparently selection is operating specifically on bill size. This sexually dimorphic grebe, by the way, takes more kinds of fishes than do more monomorphic species (Rand, 1952), probably by having effectively formed two groups of specialists, thereby increasing the size range of the prey. It also is the most gregarious grebe, hundreds of pairs nesting colonially (Nero *et al.* 1958). The relationship between niche expansion and sociality will be discussed more fully with reference to West Indian woodpeckers. It should be noted here only that in the genus *Podiceps, P. grisegena* and *P. Auritus* are solitary nesters, hold territories, and are more nearly monomorphic in bill size than is the colonial nesting *P. caspicus.*

Among lizards, some Galápagos species are sexually dimorphic, the males being larger (Berrill and Berrill, 1969), and on Grand Cayman the larger male lizards are known to take larger food than do the females (Schoener, 1967). In several species of anoles on Jamaica, the males are 2.25 times as heavy as the females, with only slightly less dimorphism being found in other anole species (Trivers, 1976). Of particular interest, however, is that the jaws are disproportionately large in these males. Also the males have stronger, stouter bodies than do females of the same weight.

Falconiformes vary greatly in their sexual dimorphism, but where it occurs, it is often "reversed," the females being larger. Amadon (1959) relates the peculiar dimorphism to their predatory foraging habits. Reversed dimorphism also occurs in other predatory birds, such as frigate birds, owls, jaegers and skuas, being most marked in bird catchers such as accipiters and falcons. Among Falconiformes sexual dimorphism is reduced in scavengers such as the New World vultures, Cathartidae (Selander, 1966), which tend to eat food of amorphous mass. It seems, then, that the size of a predator may be particularly important if its food comes in discrete sizes, a point to which we shall return.

Sexual dimorphism of island species has not generally been correlated with latitude, but Johnston and Selander (1973) found increasing sexual dimorphism with higher latitude in mainland sparrows. Their findings are consistent with Downhower's (1976), which showed that the smaller size of females in Darwin's finches was associated with first-year breeding. Small size, Downhower argued, renders the individual more sensitive to environmental changes in a fluctuating environment. Thus, small females could more quickly react to suitable breeding conditions. If the food for eggs is accumulated just before reproduction, small females are at an advantage because they can accumulate the resources necessary for egg production more quickly. (Larger females are at an advantage if the food for eggs is stored at some other time, for example, just before migration; as a result, they would have more reserves left when they reach the breeding area.)

Schoener (1965) found relatively small differences between sympatric congeners that feed on abundant food, presumably because they are able to partition the food by foraging in different microhabitats. On the other hand, he found that species exploiting less abundant food often manage to find a way to divide that food by size, and one way they may do this is by developing polymorphism in foraging structures. In addition, Schoener found that larger continental sympatric species tend to show the strongest sexual dimorphism, apparently because they feed on larger, less abundant "packets" of food. They would therefore require large foraging areas, a requirement that decreases the likelihood that niche separation can be achieved through subdivision of the habitat. Thus, increasing reliance is placed on partitioning the food by size among larger foragers. Both sexual dimorphism and the need for large foraging areas can affect social behavior, as I will show.

There is an interesting relationship between mating systems and degree of sexual dimorphism in birds. Specifically, sexual dimorphism is stronger in promiscuous and polygamous species (e.g., Tetraonidae and Icteridae). Not unexpectedly, within some families the promiscuous and polygamous species are larger than the monogamous species (Amadon, 1959). Selander (1965, 1966) argued that increased sexual dimorphism in larger species is due to stronger sexual selection in nonmonogamous breeding systems. In monogamous avian families in which the strength of sexual selection is relatively uniform among the species, the larger species are not more sexually dimorphic (in, as examples, the North American Picidae and the Corvidae). So we would expect strong sexual dimorphism to arise most frequently among nonmonogamous island colonizers. Our expectations are not met, however, in some groups, such as in certain West Indian Picidae. Such exceptions indicate the strong selective premium placed on sexual dimorphism on islands. A closer look at the relationship between the social behavior and the sexual dimorphism of certain of these species may be helpful.

First we should briefly note that presumed ancestry of some of these species. The Puerto Rican woodpecker, *M. portoricensis,* is sexually dimorphic but otherwise very similar to two North American species, the red-headed woodpecker, *M. erythrocephalus,* and the acorn woodpecker, *M. formicivorous* (Monroe, 1968; Selander and Giller, 1959, 1963). The large, sexually monomorphic Jamaican woodpecker, *C. radiolatus,* probably has, as its closest relative, the golden-fronted woodpecker, which ranges from Texas through Central America, while the strongly dimorphic Hispaniolan woodpecker, *C. striatus,* is of unknown descent.

In a study comparing the social behavior, sexual dimorphism and foraging behavior of two mainland species, the golden-fronted woodpecker and the red-bellied woodpecker, *C. carolinus,* with the Hispaniolan, Puerto Rican, and Jamaican woodpeckers, I found the two sexually dimorphic island species to be subdividing their niches according to sex, the larger-billed males hammering more frequently and gleaning more rarely than the females (Wallace, 1974b). At the same time, I noticed that the sexes remained in close association while foraging. I assumed that the close social contact was probably permitted by the reduced sexual competition that resulted from their niche partitioning. They not only foraged closer together than sexually monomorphic species, but they frequently interacted in other ways as well. For example, the Puerto Rican woodpecker formed winter foraging flocks that moved between hilltops and valleys on a daily basis. The Hispaniolan woodpecker, however, went further. The sexes not only interacted while foraging, the female closely trailing the male and frequently touching bills with him, but pairs nested colonially, as many as 26 pairs occupying one tree.

The adaptiveness of the colonial nesting has not yet been discovered, but whatever its advantage, it could not have arisen in a territorial population, so we can ask how territoriality came to be abandoned. Ashmole (1967) suggested that the Hispaniolan woodpecker lost its territorial behavior because of its unusually high density. He proposed that under high population density, territorial birds would expend too much effort defending their territories against frequent interlopers, thus unadaptively distorting their time/energy budget. If this is true, the loss of territories, in this case, can be directly associated with the development of sexual dimorphism. In an island situation, with low interspecific competition, subdivision of a niche by foraging specialization undoubtedly resulted in more efficient niche exploitation. With more food accessible, the population could have increased to the point where the maintenance of territories became uneconomical, and thus they were relinquished. At the same time, the sexual dimorphism and high population density could have interacted in a kind of positive feedback system. The increasing population density would have resulted in ever higher numbers of mating competitors, and under increasingly crowded conditions, members of pairs would have associated more closely (such as by increasing their foraging proximity) to avoid pair-bond disruption. The resulting proximity would have demanded greater niche subdivision and increased sexual dimorphism, which would have effectively increased the food supply. The feedback concept is tentatively supported by the fact that pairs remain associated the year around and that the Hispaniolan woodpecker has both the highest density and the greatest degree of sexual dimorphism of any of the species included in the study.

Whereas the coloniality of the Hispaniolan woodpecker remains unexplained, the pairs obviously now actively seek out each other's company. Grouping, however, was not necessarily adaptive for them (we don't know much about the patchiness of their food or their problems with predators). It could have been the natural consequence of nonterritorial animals' simply being attracted to their own kind.

The Hispaniolan woodpecker, then, furnishes us certain clues as to the relationship of insularity, sexual dimorphism, and social behavior. It seems that the principle of sexual dimorphism's resulting in reduced intersexual competition and increased social interaction may be generally applicable to other animals, especially vertebrates. However, the range of covariables is probably quite varied. We await a synthetic analysis of the range of social effects of sexual dimorphism for any group of animals.

E. Increase in Size

Size changes on islands are not unusual. Some species grow smaller, such as the pygmy hippos on Madagascar (Gunderson, 1976) and the dwarf elephants and deer of Mediterranean islands (Kurtén, 1968). Within some classes of animals, size shifts may be in either direction. At Ninigo, northwest of the Admiralty group, a considerable number of species are larger than their nearest mainland relatives, while at Rennell, outlying the Solomons, the endemic subspecies show a marked tendency toward reduction in size (Murphy, 1938). Islands also boast the world's largest (9 m) and the world's smallest (3 cm) lizards (Carlquist, 1965). Selection for size changes in any direction, of course, need not be ameliorated by interspecific competition on islands.

As a general rule, however, if animals change size on islands, they usually grow larger (see Carlquist, 1965). To consider a few examples, one might note the giant tortoises and iguanas of the Galápagos (Kurtén 1968), a piculet of Hispaniola, the North American ivory bill of Cuba (Lack 1976), the Guadeloupe woodpecker (unpublished data), the extinct giant dormouse of Majorca (W. Waldren, personal communication, 1976), and the finch, *G. fortis,* of the Galápagos, which increases in size on islands where its large competitor *G. magnirostris* is absent. An exhaustive compilation of size changes on islands would be most useful for the study of evolutionary ecology.

An increase in general body size would obviously result in larger trophic appendages, a shift that would be reflected in changed foraging patterns. These, in turn, could be expected to have an effect on social patterns in some cases, as I shall describe.

Increased size of trophic appendages can result, of course, not only from a change in general body size but also from selection operating specifically on body parts. The legs, toes, bills, and wings are apparently rather evolutionarily labile structures that quickly reflect minor shifts in feeding and perching habits (Keast, 1968; Wallace, 1974b). There seems to be a tendency for island birds to have larger bills than their mainland counterparts (Grant, 1965a; Keast, 1970). The longer bills of birds is often correlated with diversified feeding, a trait of niche-expanding, island-colonizing birds (Grant, 1965a; Keast, 1970).

Large size in some island colonists is obviously not a reflection of selection for large size *per se* but is instead a result of character displacement. The bullfinch *Loxigilla violacea* is larger, with a larger bill, on Jamaica, where it coexists with the "bullfinch" *Loxipasser anoxanthus,* than where it exists alone on Hispaniola and the Bahamas (Lack, 1976). So, large size on Jamaica is apparently a mechanism for avoiding competition with a smaller species. The importance of character displacement in size for island birds is reflected in the fact that a higher proportion of West Indian congeners show marked size differences than do mainland congeners (Schoener, 1965). In a sense, of course, character displacement may not only be selected for on islands but *by* islands, since only those individuals that differ to a critical degree from previous colonizing species can successfully establish themselves on an island to begin with (Grant, 1969).

Whereas there are more large-billed birds on islands than on continents at all latitudes (Volsøe, 1951; Grant, 1968), there may be a latitudinal effect on bill size. It is known, for example, that there are more large-billed insectivorous birds in the tropics than in temperate regions (Schoener and Janzen, 1968). In fact, the median exposed culmen length is 16 mm in the tropics and 11.8 mm in temperate latitudes (Schoener, 1971). Lack (1971) found that the Allen and Bergmann rules were generally operative for birds. Northern species were larger with proportionately smaller beaks, a finding that underscores the strength of the selection that produced the large tropical-island forms such as the Jamaican and Guadeloupe woodpeckers.

One reason large-billed insectivorous species are virtually confined to the tropics is that there is a much higher proportion of large insects in the tropics than in temperature regions. The large size is possibly permitted by the longer average growing season in the tropics (Schoener and Janzen, 1968). For correlations of bill size and prey size and some notable exceptions, see Schoener and Janzen (1968) and Schoener (1971).

In general, it may be assumed that taking larger food items increases foraging efficiency. In considering species of different genera, Hespenheide (1966) and Kear (1962) found that birds with long bills ate large food items more frequently and more quickly than did short-billed birds. For larger-billed birds, then, not only were food items more expeditiously handled, but the energy yield from each capture was increased. It seems that larger bill size would be most advantageous if the food items came in discrete sizes. Whereas a larger-billed flycatcher might take larger food items, increased bill size in a honey eater might have less impact on foraging efficiency.

In general, the larger size of trophic structures results in an increase in the upper limits of food sizes, as has been demonstrated in *Anolis* lizards (Schoener, 1967), sympatric species of terns (Ashmole, 1968), and finches (Pulliam, 1970). In a study of 20 temperate finch species, Pulliam found that an increase in culmen length increases the upper size limit of food without interfering with the lower limit, thus effectively increasing the *range* of food sizes. In a study of island birds, Grant (1965a) also noted that larger bills were more efficient in dealing with a larger range of food items. He suggested that the versatility was needed because of the reduced number of insect species on islands. Janzen (personal communication, 1976) has indicated that not only is species diversity generally lower on islands, but biomass within entire orders of animals is almost certainly lower than on continents. His speculation is in agreement with the "shallow-niche" model I suggested earlier, which underscores the marked advantage for any species increasing its food range. However, one problem that an island insectivore meets in increasing the range of its food sizes is the paucity of larger insects on islands. Schoener (1967) found that large insects were absent in the diets of Caribbean lizards, and Janzen (1973) found that insects over 1 cm long were absent in visual censuses on island foliage in the tropics.

Janzen found the density of large (> 10 cm snout-vent) "insectivorous" teid lizards higher on Providencia Island than in Central America. However, he also discovered that they eat a great amount of vegetable matter and that their diet is apparently associated with changes in their general behavior patterns. For example, they tend to bask in the sun, a sure invitation to predators on the mainland. The island forms are evidently not heavily preyed upon, and they need the heat gained in basking to help them quickly break down the low-energy vegetable matter in their "internal compost heap." It would be interesting to know how the sedentary basking, the switch to a localized food supply, and the loss of furtiveness have altered their social behavior.

Among birds, whereas bill size is probably not critical to any species that relies on food of amorphous quality such as fruit pulp, it may be important to frugivory if large seeds or entire fruits are eaten. Even in birds that rely exclusively on amorphous food throughout most of the year, selection may strongly influence bill size to maximize efficiency in taking other kinds of foods in the remaining period. In addition, food of discrete size may make up a small, but critical, part of the food of essentially "amorphagus" species. For example, nearly all species of tanagers rely largely on the same species of fruit, but each takes its own kind of alternative insect prey (Snow and Snow, 1971). In addition, Lack (1976) found much greater food-choice overlap among frugivorous birds than among insectivorous forms. It seems, then, that interspecific flocking of frugivorous species would increase the probability of species competitiveness, but correlations of diet and social behavior have not yet been extensively investigated.

It has been noted that there are more frugivorous bird species on islands than on continents (Lack, 1976), but there is also an important latitudinal effect on diet and consequent social behavior. In north temperate regions, normally segregated species may congregate to exploit temporarily superabundant food (Lack, 1946). But in the tropics, fruits are available throughout the year and so more bird species have become fruit specialists, with a resulting reduction in species that gather at a common food supply.

Perhaps the influence of large size on the social behavior of an island species can best be illustrated by a consideration of the Jamaican woodpecker, *C. radiolatus*. This species is sexually monomorphic in bill size (Table I) and in general body size, as indicated by tarsus length (Table II). Tarsus length is used here as an index of body size since body weights were not available in every case (but see Grant, 1966c). The Jamaican woodpecker is one of the larger species of Antillean melanerpines, the largest being *M. superciliaris* of Jamaica (L. Short, personal communication, 1974), followed by the Guadeloupe woodpecker, *M. herminieri* (Bond, 1971). Whereas the latter species shares the upper limits in size with the Jamaican woodpecker, it is sexually dimorphic both in bill size and in body size, the females being significantly smaller than the males.

In field studies of four species, I determined that those species that are more sexually dimorphic in bill size tend to a stronger subdivision of the niche by sex (Table III). In addition, the greatest foraging proximity, in terms of foraging height, occur in the more dimorphic sexes. So, the sexes of the monomorphic Jamaican woodpecker forage in similar ways but in

Table I. Bill Length

	Males				Females					
	$\bar{X}$	Range	SD	n	$\bar{X}$	Range	SD	n	$1/s^a$	t
Centurus aurifrons[b]	24.1	21.1–27.6	1.08	13	23.9	21.9–26.5	0.94	12	1.01	—
C. carolinus[b,c]	24.2	20.9–27.3	1.25	48	22.0	20.0–23.7	1.19	28	1.10	[d]
Melanerpes portoricensis[b,e]	22.7	20.3–23.4	1.18	7	18.3	17.2–20.2	0.99	8	1.24	[d]
C. striatus[b,e]	26.7	24.0–28.1	1.15	22	20.9	18.8–23.7	1.18	28	1.28	[d]
M. herminieri[e]	28.5	27.7–29.8	0.80	6	22.9	21.6–24.1	0.93	6	1.24	[d]
C. radiolatus[e]	27.1	24.5–29.0	1.45	13	26.3	25.0–29.1	1.42	8	1.03	—

[a] Ratio of larger mean measurement to smaller, for the sexes.

[b] The morphology, foraging behavior, and social behavior of these species are more completely discussed in Wallace (1969, 1974). These data are included for comparative purposes.

[c] *C. carolinus* is found to be sexually monomorphic when samples are taken throughout its range (Selander, 1966; Ridgeway, 1911). The dimorphism at the eastern border of its range, as shown here, may be due to the ecological pressures in fringing areas. Also, the sexes in this sample fail to reach 90% joint nonoverlap, a value considered ecologically important (Mayr, Linsley, and Usinger, 1953).

[d] Significant sexual difference according to t-test analysis; $p < 0.02$.

[e] Antillean species.

Table II. Tarsus Length

	Males				Females					
	$\bar{X}$	Range	SD	n	$\bar{X}$	Range	SD	n	$1/s^a$	t
Centurus aurifrons	23.1	21.0–23.8	0.60	13	22.9	19.9–23.3	0.99	12	1.01	—
C. carolinus	21.6	20.0–24.5	0.84	48	20.5	17.8–22.3	1.08	28	1.05	b
Melanerpes portoricensis[c]	21.7	19.3–23.0	1.22	7	19.7	18.6–20.9	0.81	7	1.10	—
C. striatus[c]	23.4	21.1–25.4	1.06	23	21.5	20.0–22.4	0.78	29	1.09	b
M. herminieri[c]	25.7	23.5–27.2	1.40	7	22.6	21.0–24.4	1.40	6	1.14	b
C. radiolatus[c]	24.3	23.4–26.3	0.74	13	24.8	21.9–27.0	1.42	8	1.08	—

[a] Ratio of larger mean measurement to smaller, for the sexes.
[b] Significant sexual difference according to t-test analysis; $p < 0.02$.
[c] Antillean species.

different places, whereas the sexually dimorphic species employ different foraging modes but forage in close proximity. Apparently, since the sexes of the Jamaican woodpecker do not differ in bill size, they tend to take the same food, and the resulting intersexual competition demands that they forage farther apart. The sexes of the more dimorphic forms, on the other hand, compete less and can forage closer together. The relationship of foraging-mode overlap and foraging-height overlap is shown in Fig. 2.

In the Jamaican woodpecker, both sexes are large and it may be that large size has developed as an *alternative* to sexual dimorphism. The advantages of such a strategy would be twofold. First, with larger trophic

Table III. Foraging Mode

	Probing[a]				Gleaning				Hammering			
	Males		Females		Males		Females		Males		Females	
	n^b	$\%^c$	n	$\%$	n	$\%$	n	$\%$	n	$\%$	n	$\%$	χ^2
Centurus carolinus[d]	48	(40)	31	(36)	47	(39)	43	(51)	24	(20)	11	(13)	—
Melanerpes portoricensis[e]	50	(29)	27	(24)	46	(26)	49	(43)	78	(45)	38	(33)	f
C. striatus[e]	89	(36)	110	(23)	100	(41)	281	(60)	56	(23)	80	(17)	f
C. radiolatus[e]	17	(16)	17	(17)	76	(72)	67	(67)	13	(12)	16	(16)	—

[a] Probing: inserting the bill and/or tongue into a crevice to remove prey; gleaning: picking up food items from the surface of a plant; hammering: chipping away wood or bark to expose burrowed prey. Other foraging techniques occurred too infrequently to permit statistical analysis.
[b] Number of records in the category.
[c] Percentage of total records for the sex.
[d] Number of records in the category.
[e] Antillean species.
[f] Significant sexual differences according to χ^2 analysis; $p < 0.05$.

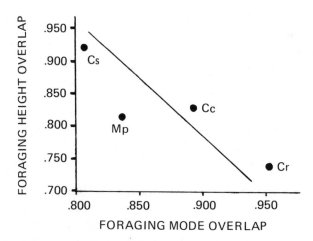

Fig. 2. Overlap in foraging mode plotted against overlap in foraging height, for the sexes. Cs—*Centurus striatus;* Cc—*C. carolinus;* Cr—*C. radiolatus;* Mp—*Melanerpes portoricensis.* Overlap is determined by Schoener's (1968) formula:

$$D = 1 - \tfrac{1}{2} \sum_{i=1}^{n} |P_{x,i} - P_{y,i}| \text{ where } P_{x,i} \text{ and } P_{y,i}$$

are the frequencies for sexes x and y, respectively, for the ith category.

appendages, it should be possible to increase the prey size, larger prey yielding more available energy. (The maintenance of the predator's larger general body size would diminish this advantage somewhat.) However, a greater advantage of large trophic appendages may be in permitting the birds to take a greater range of food types. In the case of the Jamaican woodpecker, an increase in bill size would undoubtedly add food items from the larger end of the size spectrum without seriously interfering with the ability of the birds to take the smaller food items.

As increase in the range of food sizes would result in an increase in the number of food species, especially for woodpeckers, which rely heavily on foods of discrete sizes, such as insects, seeds, and berries. I believe that the great advantage of increased bill size may be in increasing the number of potential food *species*. This increase would be particularly important in tropical and neotropical areas, where populations of different insect and plant species might be expected to peak at different times throughout the year, where the reproductive period may be extended for each population, and where harsh density-independent factors do not operate to reduce the populations of many food species simultaneously. With more food species available throughout the year, then, the Jamaican woodpecker could exploit a number of food types and switch from one to another as each came into production.

If increased bill size does confer foraging advantages, and if these advantages differ from those associated with sexual dimorphism in bill size, might we then ask how the two different strategies could have arisen? I suggest that the differences may spring from differences in the social patterns of the parent stock. Whereas the Jamaican woodpecker seems to spend relatively little time in sexual interaction, the sexes of the Hispaniolan and Puerto Rican species frequently interact, the males continually exhibiting their dominance over the females. For example, in a pair of foraging Hispaniolan woodpeckers, the male usually initiates the flights to new foraging sites, the female following closely behind. The female often lands so near the male that they can touch each other, the brief sparring with the bill ensues, after which the female usually gives way. Also, the males often supplant females that have found food. The female readily gives up her foraging site but generally continues her search near the male. Thus, the pair remains in close association. Strong sexual dimorphism in bill size could thus permit foraging proximity by allowing the sexes to search for different food, or the same food in different ways, while maintaining close social contact (Kilham, 1965; Ligon, 1973).

Carlquist (1965) described this sort of relationship between size changes and social behavior among lizards. He first noted that gigantism may not be tied merely to conversion to a new diet but to an increase in the *kinds* of foods. He suggested that a larger species of lizard may be able to eat larger food items than its smaller ancestors, while retaining the ability to eat smaller food. In addition, he argued that gigantism in reptiles may be related to territorialism. A larger animal, he noted, could police a larger territory and thus ensure a larger food supply.

We cannot determine, at present, what may be responsible for the presumed social differences in the Jamaican woodpecker and its more gregarious cousins. Perhaps the answer lies in differences in predation, present or historical—stronger predation selecting for disassociation of pairs as a means of avoiding detection. Or there may be differences in food distribution between Jamaica and the other islands that have not yet been discovered. And there remains the possibility that the social systems have not developed on any adaptive basis at all, but this should be our explanation of last resort.

XII. SUMMARY

Islands may offer novel environments to their colonizers, but the offering has strings attached. It comes with its own stringent demands. Whereas colonizers may find few competitors, they may also find conditions so different from these they left that they must rapidly adapt to their new world or perish. That adaptation can be of several types. If the individuals are behaviorally flexible, they may adapt on an individual basis by drawing on new segments of their behavioral repertoire. If they have a short generation time and a high genetic load, they may more quickly adapt on a populational basis. They have a greater period to effect any change if the island presents them with a benign environment, or if they are preadapted through evolving in a similar environment. The adaptive changes, on whatever basis, have traditionally been investigated individually: How have foraging patterns changed? What size changes are involved? Have niches expanded? How have social systems changed? It should be apparent, though, that the conceptual segregation of the components of an adaptive complex is unrealistic and simplistic and, worse, can lead to unproductive investigative paradigms, since selection cannot be assumed to operate on individual phenotypic traits while holding others constant.

Even when the adaptive syndrome is considered, the relationships often assume a peculiar linearity. For example, it has been assumed that sexual selection can influence size of trophic appendages, with a resulting influence on foraging behavior. However, such explanatory efforts answer only a small part of a larger question. We must now begin to consider more fully the complex feedback relationships between morphology, foraging behavior, and social systems in different kinds of habitats, islands providing us with simpler places to test our ideas. The three factors:

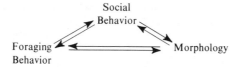

are not independent variables. Changes in one can be expected to alter the others as the environment winnows sensitive cytoplasm, producing ever-changing morphological and behavioral phenotypes through complex feedback systems. One can ask how a certain system may have been initiated or how strong the selection may be in some direction under certain conditions, but behind such questions must be a clear awareness of the reciprocal nature of the system.

XIII. REFERENCES

Amadon, D. (1953). Avian systematics and evolution in the Gulf of Guinea. *Bull. Am. Mus. Nat. Hist.* **100**:397–431.

Amadon, D. (1959). The significance of sexual differences in size among birds. *Proc. Am. Phil. Soc.* **103**:531–536.

Armstrong, J. T. (1965). Breeding home range in the Nighthawk and other birds: Its evolutionary and ecological significance. *Ecology* **46**:619–629.

Ashmole, N. P. (1961). *The Biology of Certain Terns*, Ph.D. dissertation, Oxford University.

Ashmole, N. P. (1967). Sexual dimorphism and colonial breeding in the woodpecker, *Centurus striatus*. *Am. Nat.* **101**:353–356.

Ashmole, N. P. (1968). Body size, prey size and ecological segregation in five sympatric tropical terns (Aves: Laridae). *Syst. Zool.* **17**:292–304.

Berrill, N. J., and Berrill, M. (1969). *The Life of Sea Islands*, McGraw-Hill, New York.

Bond, J. (1971). *Birds of the West Indies*, Houghton Mifflin, Boston.

Brown, W. L., and Wilson, E. O. (1956). Character displacement. *Syst. Zool.* **5**:49–64.

Carlquist, S. (1965). *Island Life*, Natural History Press, Garden City, N.Y.

Case, T. J. (1975). Species numbers, density compensation, and the colonizing ability of lizards on islands in the Gulf of California. *Ecology* **56**:3–18.

Crook, J. H. (1961). The todies (Ploceinae) of the Seychelles Islands. *Ibis* **103a**:517–548.

Crook, J. H., and Gartlan, J. S. (1966). Evolution of primate societies. *Nature* **210**:1200–1203.

Crowell, K. L. (1962). Reduced interspecific competition among the birds of Bermuda. *Ecology* **43**:75–88.

Darlington, P. J. (1970). Carabidae on tropical islands, especially the West Indies. *Biotropica* **2**:7–15.

Darwin, C. (1859). *The Origin of Species*, republished 1928, Dutton, New York.

Diamond, J. M. (1970). Ecological consequences of island colonization by Southwest Pacific birds. I: Types of niche shifts. *Proc. Nat. Acad. Sci.* **67**:529–536.

Diamond, J. M. (1972). Avifauna of the Eastern Highlands of New Guinea. *Publ. Nuttall Orn. Club*, No. 1, Cambridge, Mass.

Dobzhansky, T. (1961). On the dynamics of chromosomal polymorphism in *Drosophila*. In Kennedy, J. S. (ed.), *Insect Polymorphism, Symp. Royal Ent. Soc.*, No. 1:30–42.

Dobzhansky, T. (1963). Rigid vs. flexible chromosomal polymorphisms in *Drosophila*. *Am. Nat.* **96**:321–328.

Dobzhansky, T. (1965). Genetic diversity and fitness. In *Genetics Today*, Vol. 3, *Proc. XI Intern. Congr. Genetics*, pp. 541–552.

Downhower, J. F. (1976). Darwin's finches and the evolution of sexual dimorphism in body size. *Nature* **263**:558–563.

Grant, P. R. (1965a). The adaptive significance of some size trends in island birds. *Evolution* **10**:355–367.

Grant, P. R. (1965b). Plumage and the evolution of birds on islands. *Syst. Zool.* **14**:47–52.

Grant, P. R. (1965c). A systematic study of the terrestrial birds of the Tres Marías Islands, Mexico. Postilla, *Yale Peabody Mus. Nat. Hist.* **90**:1–106.

Grant, P. R. (1966a). The coexistence of two wren species of the Genus *Thryothorus*. *Wilson Bull.* **78**:266–278.

Grant, P. R. (1966b). The density of land birds on the Tres Marías Islands in Mexico. I: Numbers and biomass. *Can. J. Zool.* **44**:391–400.

Grant, P. R. (1966c). Ecological compatibility of bird species on islands. *Am. Nat.* **100**:451–462.

Grant, P. R. (1967). Bill length variability in birds of the Tres Marías Islands, Mexico, *Can. J. Zool.* **45**:805–815.

Grant, P. R. (1968). Bill size, body size, and the ecological adaptations of bird species to competitive situations on islands. *Syst. Zool.* **17**:319–333.

Grant, P. R. (1969). Community diversity and the coexistence of congeners. *Am. Nat.* **103**:552–556.

Gunderson, H. L. (1976). *Mammalogy*, McGraw-Hill, New York.

Hamilton, W. D. (1964). The genetical evolution of social behavior. I, II. *J. Theor. Biol.* **7**:1–52.

Hespenheide, H. A. (1966). The selection of seed size by finches. *Wilson Bull.* **78**:191–197.

Janzen, D. H. (1973). Sweep samples of tropical foliage insects: Effects of seasons, vegetation types, elevation, time of day, and insularity. *Ecology* **54**:687–708.

Janzen, D. H., and Schoener, T. W. (1968). Differences in insect abundance and diversity between wetter and drier sites during a tropical dry season. *Ecology* **49**:96–110.

Johnston, R. F., and Selander, R. K. (1973). Evolution in the house sparrow. III: Variation in size and sexual dimorphism in Europe and North and South America. *Am. Nat.* **107**:373–390.

Karr, J. R. (1971). Structure of avian communities in selected Panama and Illinois habitats. *Ecol. Monogr.* **41**:207–233.

Kear, J. (1962). Food selection in finches, with special reference to interspecific differences. *Proc. Zool. Soc. Lond.* **138**:163–204.

Keast, A. (1968). Competitive interactions and the evolution of ecological niches as illustrated by the Australian honey-eater genus *Melithreptus* (Meliphagidae). *Evolution* **22**:762–784.

Keast, A. (1970). Adaptive evolution and shifts in niche occupation in island birds. *Biotropica* **2**:61–75.

Kilham, L. (1965). Differences in feeding behavior of male and female hairy woodpeckers. *Wilson Bull.* **77**:134–145.

Klopfer, P. H. (1959). Environmental determinants of faunal diversity. *Am. Nat.* **93**:337–342.

Klopfer, P. H., and MacArthur, R. H. (1961). On the causes of tropical species diversity: Niche overlap. *Am. Nat.* **95**:223–226.

Krebs, C. J. (1972). *Ecology, the Experimental Analysis of Distribution and Abundance*, Harper and Row, New York.

Krebs, C. J., Keller, B., and Tamarin, R. (1969). *Microtus* population biology. *Ecology* **50**:587–607.

Kurtén, B. (1968). *Pleistocene Mammals of Europe*, Weidenfeld and Nicolson, London.

Lack, D. (1944). Ecological aspects of species-formation in passerine birds. *Ibis* **86**:260–286.

Lack, D. (1946). Competition for food by birds of prey. *J. Anim. Ecol.* **15**:123–129.

Lack, D. (1947). *Darwin's Finches*, University Press, Cambridge.

Lack, D. (1970). Island birds. *Biotropica* **2**:29–31.

Lack, D. (1971). *Ecological Isolation in Birds*, Harvard Press, Cambridge, Mass.

Lack, D. (1976). *Island Biology, Illustrated by the Land Birds of Jamaica*, University of California Press, Berkeley.

Lack, D., and Southern, H. N. (1949). Birds of Tenerife. *Ibis* **91**:607–626.

Ligon, J. D. (1973). Foraging behavior of the white-headed woodpecker in Idaho. *Auk* **90**:862–869.

MacArthur, R. H. (1969). Patterns of communities in the tropics. *J. Linn. Soc. Lond. Biol.* **1**:19–30.

MacArthur, R. H. (1972). *Geographical Ecology*, Harper and Row, New York.

MacArthur, R. H., and Pianka, E. (1966). On optimal use of a patchy environment. *Am. Nat.* **100**:603–609.

MacArthur, R. H., and Wilson, E. O. (1963). An equilibrium theory of insular zoogeography. *Evolution* **17**:373–387.

MacArthur, R. H., and Wilson, E. O. (1967). *The Theory of Island Biogeography*, Princeton University Press, Princeton.

MacArthur, J., Diamond, J., and Karr, J. (1972). Density compensation in island faunas. *Ecology* **53**:330–342.

MacArthur, R. H., MacArthur, J., MacArthur, D., and MacArthur, A. (1973). The effect of island area on population densities. *Ecology* **54**:657–658.

Marler, P., and Boatman, D. J. (1951). Observations on the birds of Pico, Azores. *Ibis.* **93**:90–99.

Matthews, R. W., and Matthews, J. R. (1970). Adaptive aspects of insular evolution: a symposium. *Biotropica* **2**:1–2.

Maynard Smith, J. (1971). What use is sex? *J. Theor. Biol.* **30**:319–335.

Mayr, E. (1942). *Systematics and the Origin of Species,* Columbia University Press, New York.

Mayr, E. (1963). *Animal Species and Evolution,* Harvard University Press, Cambridge.

Mayr, E., Linsley, E. G., and Usinger, R. L. (1953). *Methods and Principles of Systematic Zoology,* McGraw-Hill, New York.

Monroe, B. L. (1968). A distributional survey of the birds of Honduras. *Ornithol. Monogr.* No. 7 (A.O.U.).

Moreau, R. E. (1966). *The Bird Faunas of Africa and Its Islands,* Academic Press, New York.

Moreau, R. E., and Ridpath, M. G. (1966). The birds of Tasmania: Ecology and evolution. *Ibis* **108**:348–393.

Murphy, R. C. (1938). The need of insular exploration as illustrated by birds. *Science* **88**:533–539.

Nero, R. W., Lahrman, F. W., and Bard, F. G. (1958). Dry-land nest-site of a western grebe colony. *Auk* **75**:347–349.

Orians, G. H. (1969). The number of bird species in some tropical forests. *Ecology* **50**:783–801.

Parkes, K. C. (1965). Character displacement in some Philippine cuckoos. *The Living Bird, 4th Annual Report of the Cornell Laboratory of Ornithology,* pp. 89–98.

Power, D. M. (1972). Number of bird species on the California Islands. *Evolution* **26**:451–463.

Preston, F. W. (1962). The canonical distribution of commoness and rarity. I, II. *Ecology* **43**:185–215, 410–432.

Pulliam, H. R. (1970). *Comparative Feeding Ecology of a Tropical Grassland Finch* (*Tiaris olivacea*), Ph.D. dissertation, Duke University.

Pulliam, H. R. (1973). Comparative feeding ecology of a tropical grassland finch (*Tiaris olivacea*). *Ecology* **54**:284–299.

Rand, A. L. (1952). Secondary sexual characters and ecological competition. *Fieldiana-Zoology* **34**:65–70.

Ridgeway, R. (1911). The birds of North and Middle America. Part 4. *Bull. U.S. Nat. Mus.* **50**:1–973.

Ripley, S. D., and Bond, G. M. (1966). The birds of Socotra and Abd-El-Kuri. *Smithson. Misc. Collect.* **151**:1–37.

Schoener, T. W. (1965). The evolution of size differences among sympatric congeneric species of birds. *Evolution* **19**:189–213.

Schoener, T. W. (1967). The ecological significance of sexual dimorphism in size in the lizard *Anolis conspersus. Science* **155**:474–476.

Schoener, T. W. (1971). Large-billed insectivorous birds: A precipitous diversity gradient. *Condor* **73**:154–161.

Schoener, T. W., and Janzen, D. H. (1968). Some notes on tropical versus temperate insect size patterns. *Am. Nat.* **102**:207–224.

Selander, R. K. (1965). On mating systems and sexual selection. *Am. Nat.* **99**:129–141.

Selander, R. K. (1966). Sexual dimorphism and differential niche utilization in birds. *Condor* **68**:113–151.

Selander, R. K., and Giller, D. R. (1959). Interspecific relations of woodpeckers in Texas. *Wilson Bull.* **71**:107–124.

Selander, R. K., and Giller, D. R. (1963). Species limits in the woodpecker genus *Centurus* (Aves). *Bull. Am. Mus. Nat. Hist.* **124**:213–274.

Selander, R. K., Yang, S. Y., and Hunt, W. G. (1969). Polymorphism in esterases and hemoglobin in wild populations of the house mouse (*Mus musculus*). *Studies in Genetics V*, pp. 271–338.

Selander, R. K., Smith, M. H., Yang, S. Y., Johnson, W. E., and Gentry, J. B. (1971). Biochemical polymorphisms and systematics in the genus Peromyscus. I: Variations in the old-field mouse. *Studies in Genetics VI*, Texas Univ. Publ. **7103**:49–90.

Serventy, D. L. (1951). Inter-specific competition on small islands. *West Austr. Nat.* **3**:59–60.

Sibley, C. G. (1957). The evolutionary and taxonomic significance of sexual dimorphism and hybridization in birds. *Condor* **59**:166–191.

Simberloff, D. (1976). Experimental zoogeography of islands: Effects of island size. *Ecology* **57**:629–648.

Snow, B. K., and Snow, D. W. (1971). The feeding ecology of tanagers and honeycreepers in Trinidad. *Auk* **88**:291–322.

Soulé, M. (1972). Phenetics of natural populations. III: Variation in insular populations of a lizard. *Am. Nat.* **106**:429–446.

Soulé, M., and Yang, S. Y. (1973). Genetic variation in side-blotched lizards on islands in the Gulf of California. *Evolution* **27**:593–600.

Terborgh, J. (1973). Chance, habitat and dispersal in the distribution of birds in the West Indies. *Evolution* **27**:338–349.

Thielcke, G. (1965). Gesangsgeographische Vestion des Gartenbaumläufers (*Gerthia brachydactyla*) in Hinblick auf das Artbildungsproblem. *Z. Tierpsychol.* **22**:542–566.

Trivers, R. (1976). Sexual selection and resource-accruing abilities in *Anolis garmani*. *Evolution* **30**:253–269.

Van Valen, L. (1965). Morphological variation and width of ecological niche. *Am. Nat.* **99**:377–390.

Volsøe, H. (1951). The breeding birds of the Canary Islands. I: Introduction and synopsis of the species. *Vidensk. Medd. Dan. Naturhist. Foren. Kbh.* **113**:1–153.

Wallace, A. R. (1880). *Island Life; or the Phenomena and Causes of Insular Faunas and Floras, Including a Revision and Attempted Solution of the Problem of Geological Climates*, Macmillan, London.

Wallace, A. R. (1895). *Natural Selection and Tropical Nature*, Macmillan, London.

Wallace, R. A. (1969). *Sexual Dimorphism, Niche Utilization, and Social Behavior in Insular Species of Woodpeckers*, Ph.D. dissertation, University of Texas, Austin.

Wallace, R. A. (1973). *The Ecology and Evolution of Animal Behavior*, Goodyear, Santa Monica.

Wallace, R. A. (1974a). Aberrations in the tongue structure of some melanerpine woodpeckers. *Wilson Bull.* **86**:79–82.

Wallace, R. A. (1974b). Ecological and social implications of sexual dimorphism in five melanerpine woodpeckers. *Condor* **76**:238–248.

Watson, G. (1964). Ecology and evolution of passerine birds on the islands of the Aegean Sea. *Diss. Abstr. B Sci. Eng.* 1242.

Williams, E. E. (1969). The ecology of colonization as seen in the zoogeography of anoline lizards on small islands. *Q. Rev. Biol.* **44**:345–389.

Williams, G. C. (1975). *Sex and Evolution*, Princeton University Press, Princeton.

Wilson, E. O. (1959). Adaptive shift and dispersal in a tropical ant fauna. *Evolution* **13**:122–144.

Wilson, E. O. (1961). The nature of the taxon cycle in the Melanesian ant fauna. *Am. Nat.* **95**:169–193.

Wilson, E. O. (1965). The challenge from related species. In Baker, H. G., and Stebbins, G. L. (eds.), *The Genetics of Colonizing Species,* Academic Press, New York, pp. 7–27.

Yarbrough, C. G. (1970). Summer lipid levels of some subarctic birds. *Auk* **87**:100–110.

Zahavi, A. (1975). Mate selection—A selection for a handicap. *J. Theor. Biol.* **53**:205–214.

Chapter 9

ON PREDATION, COMPETITION, AND THE ADVANTAGES OF GROUP LIVING

Daniel I. Rubenstein[1]

Department of Zoology
Duke University
Durham, North Carolina 27706

I. ABSTRACT

Many evolutionary reasons have been suggested as to why animals live in groups. Based on the costs and benefits associated with group living it appears that the risk of predation and the competitively induced need to forage efficiently are the most important forces responsible for the formation and maintenance of groups.

Most field studies and mathematical models that have investigated the environmental conditions that favor the formation of groups have studied the effects of either predation or competition. After examining some of these studies and their major conclusions, I present a model, based on the theory of games, that focuses on the combined action of predation and competition. Simulations under a variety of environmental conditions reveal that individuals that are poor competitors are more likely to remain in groups when the visibility of the habitat is low, when the level of competitive inequality among group members is low, when the carrying capacity of the habitat is high, and when the presence of neighbors enhances an individual's feeding success.

Models, such as this one, that incorporate the effects of predation and resource competition, while accounting for how differences in age, sex, size, and past experience influence competitive ability, provide insights into the dynamics of social behavior. When other factors, such as kinship relationships and subtle interspecific interactions, are also considered, a general and predictive theory of social organization can be developed.

[1] Present Address: King's College Research Centre, Cambridge CB2 1ST, England, U.K.

II. INTRODUCTION

One of the most studied and most poorly understood concerns of ethology is why animals live in groups. Living in groups has been claimed to aid in the rearing of young, to facilitate mating, to increase foraging success, to reduce the risk of predation, to provide protection from inclement weather, and to increase swimming efficiency. Undoubtedly each of these advantages has contributed to the formation of groups in particular situations. Nevertheless, constructing a simple patchwork of advantages that can be used to justify the existence of groups often differing with respect to habitat and the sex, age, and degree of relatedness of its constituents provides no real understanding of why animals live in groups. If a predictive theory of social organization is to emerge, it will require an understanding of how various selective pressures operate so that an individual's chances of survival or reproductive success are greater when it is living in a particular type of group than when it is living alone or in a different type of group. In particular, ethologists must (1) determine which are the most important selective forces; (2) determine how the action of each force is influenced by characteristics of the environment, attributes of the individual, and species-specific constraints; and (3) determine the ramifications of the combined action of these selective forces. From such an evolutionary understanding of social dynamics, a predictive theory of social organization can be developed.

The purpose of this paper is to examine how much progress has been made toward the achievement of each of the three objectives. Pausing to reflect upon what has already been accomplished will provide insights into what still needs to be investigated.

III. THE MAJOR SELECTIVE FORCES

It is easy to speculate on how natural selection operates, but proving how it operates is one of the most difficult tasks in evolutionary biology. Even if one assumes that a behavior is beneficial and has been favored by natural selection, it is very difficult to prove that selection has favored the behavior because of a particular consequence.

For example, black-headed gulls remove broken eggshells from their nests. Tinbergen and his co-workers (1962) suggested that selection might favor the removal of broken eggshells from the nest because such behavior might reduce the predator's ability to locate the nest, reduce the chances of a chick's cutting itself on sharp edges, reduce the growth of bacteria, or increase the parents' ability to brood the young. Although these authors performed a series of experiments demonstrating that eggshell removal served to maintain the camouflage of the nest, others (e.g., McFarland, 1976) have contended that it is still possible that the other considerations might also exert small selection pressures.

Studying the adaptive value of behavior, such as group living, is even more difficult when experimental manipulations are not possible. Probably the most realistic approach involves listing possible advantages and disadvantages associated with the behavior, then noting the frequencies at which the benefits associated with the advantages and the costs associated with the disadvantages occur. Based on the distribution of costs and the occurrence of benefits, an understanding of the important selective forces and how they operate will emerge.

Using this approach, Alexander (1974) proposed that groups form only when the advantages derived from the presence of others—reduced risk of predation and enhanced feeding efficiency—offset the detriments of group living: automatically intensified competition and increased disease and parasite transmission, as well as potentially increased chances of injury because of competition, increased conspicuousness to predators, and increased chances of misdirected parental care. Based on this statement, predation and the competitively induced need to forage efficiently should be the two major selection pressures that determine whether an individual lives alone or in a group.

Unfortunately, the paucity of field studies measuring both the cost and the benefits of group living makes it difficult to verify that predation and competition are the two major forces responsible for the formation of groups. In one study on colonial nesting bank swallows, Hoogland and Sherman (1976) uncovered several detrimental aspects of group living: increased competition for nest sites, nest material, and mates; increased likelihood of misdirected parental care; and increased parasite transmission. These authors concluded, however, that there was only one advantage of colonial living: predator defense. They noted that increases in colony size resulted in more rapid detection, more vocal harassment, and greater percentages of inhabitants participating in the mobbing of predators. In a few instances, this mobbing resulted in the predator's being driven away. Although Hoogland and Sherman found no evidence that coloniality enhanced feeding success, Emlen (1975) observed that bank swallows of another colony left in groups to forage and suggested that this social foraging led to increased reproductive success.

Even though this evidence by no means implies that predation and competition are the only important selective pressures responsible for the formation of groups, it does suggest that they do play a major role. The fact that in one area it appears that predation was the chief pressure for bank swallow coloniality whereas in another area it appears that feeding demands were responsible underscores the need for more comparative studies, especially those that compare similar groups under different environmental conditions.

Theoretical studies also play an important role in determining how the action of predation and competition influences the formation of groups. If characteristics of the environment, the species, and the individuals are incorporated into models of predatory and competitive behavior, the actions most beneficial to predators and prey can be predicted. For example, Treisman (1975a) used models that incorporated the perceptual abilities of the predator and the prey to predict under what environmental conditions the prey should live in groups. But if a realistic and predictive theory of social organization is to be achieved, both empirical and theoretical studies must be pursued simultaneously. The models provide an understanding of how predation and competition might operate and thus help focus field studies. Therefore, the remainder of this paper focuses on theoretical studies. Nevertheless, empirical studies provide basic biological data that, by making the assumptions underlying the models realistic, help ensure that the predictions will be biologically meaningful.

IV. PREDATION AND RESOURCE COMPETITION: INFLUENCES OF ENVIRONMENT, INDIVIDUALS, AND SPECIES

A. Predation

Animals can reduce the risk of predation by increasing their probability of detecting the predator and responding—by concealment, avoidance, or defense—before the predator can detect and attack them. Thus, we might ask: Are animals in groups more likely to be successful at reducing the risk of predation than are solitary animals? And if so, under what circumstances?

Based on some simple assumptions, Pulliam (1973) demonstrated theoretically that birds in flocks could benefit from the scanning behavior of neighbors. By assuming that birds cock their heads to scan for predators, that there is a mean head-cocking rate, and that the attacking predators are exposed for some mean time period, he showed that birds in a group are more likely than solitary birds to detect a predator and fly to safety before being killed. In addition, he showed that birds in flocks could derive the same level of predator protection as do solitary birds, even if each individual lowered its level of vigilance.

Even if animals are randomly assembled when a predator appears, Hamilton (1971) has shown that the formation of a group offers immediate benefits. His simplest hypothetical example depicts a snake emerging from a pond about which frogs are distributed at random. If the predator were to strike the nearest frog, Hamilton showed that a frog could reduce its chances of being the nearest to the predator by jumping into the gap between two other frogs. As every frog performed these calisthenics, the arrangement would become less random, and eventually a densely packed aggregation would be produced. (Similar conclusions were reached when the geometric model was expanded to include animals living on the plain.) One of the elegant features of the model is that the formation of a group requires only that animals behave in their own self-interest.

Pulliam (1973) has claimed, however, that purely selfish behavior does not lead to the formation of an aggregation. In spite of all the realignments that take place, some individuals still remain on the periphery, subject to a greater risk of predation than those in the interior. Pulliam implicitly assumed that because the risk of predation on the periphery is greater than that of living alone, the outermost individuals should abandon the group. And because each successive inner layer eventually becomes the peripheral layer, the group should eventually deteriorate. Thus, unless the risk of predation for the peripheral individuals can be lowered, Pulliam argued that groups should not form.

Pulliam was certainly correct in elaborating on how Hamilton's scheme accentuates individual inequalities. But although an individual on the outside of a group suffers the greatest risk of predation during any particular encounter with the predator, it is not valid to assume that over a number of such encounters an individual's average risk is automatically higher than the risk suffered by animals living alone. If the same individuals always acquired the central spots, then the average risk to the peripheral individuals probably would be higher than that of a solitary individual. Thus, groups would not form unless the peripheral individuals received ancillary benefits. But if all the individuals had an equal chance of being in the center, then the average predation risk might be lower for peripheral individuals than for animals living alone. In this case, then, simple geometric considerations would lead to the formation of a group.

In certain respects, it does not really matter if Hamilton's theory of the selfish herd is sufficient to account for the formation of groups. According to Alexander (1974), behavior providing ancillary benefits that make group living even more attractive should subsequently evolve for two reasons. First, social behavior may evolve to minimize the disadvantages inherent in group living. For example, social grooming in mammals might have evolved to reduce parasite transmission; communicatory and dominance systems might have evolved to limit the intensity of competition. Second, social behavior may evolve to enhance the original advantage that was responsible for the formation of the group. Thus, if a group formed as a means of reducing each individual's risk of being preyed upon, it is likely that additional types of behavior that further reduce this risk, such as collective predator detection, would be favored by natural selection.

The major weakness of Hamilton's idea of the selfish herd is that it is based on the assumption that the predator instantly appears among the prey. Probably very few predators are able to do so. More likely, they wander about the habitat detecting and attacking prey from outside the perimeter of the population. Therefore, Vine (1971) believed that in order to determine how effective different arrangements of prey are at reducing the risk of predation, one must consider the perceptual abilities of the predator. His model is based on the idea that the predator's ability to detect prey visually increases as the magnitude of the prey's minimum dimension increases. Because he considered height to be an ungulate's minimum dimension, he concluded that a group would be no easier to perceive than solitary individuals. But because animals in a group are localized in one point in space, the predator would, on average, have to turn its head through a greater angle to detect prey in a group than to detect those randomly scattered about the environment. Thus, it would take longer for a predator to locate animals in a group. If this mean time to detection exceeded the time allotted by the predator to scanning the environment, the group would escape being preyed upon. Thus, Vine concluded, group living could be advantageous, because groups are difficult for the predator to detect.

Although Vine's conceptualization of the problem has influenced many other studies, his model suffers from some unrealistic assumptions about predators' foraging habits and the mechanisms of visual perception. Most predators probably do not leave an area when no prey are detected during a single scan of the environment. As Treisman (1975a) has indicated, animals often possess territories or home ranges that provide them with information about the habitat and habits of the prey. Because of this familiarity, they probably do not leave an area after one negative scan. Instead, Treisman suggested that predators stay in an area until the cost of staying and rescanning the habitat exceeds the cost of leaving. The actual number of rescans that occur before the habitat is deemed unprofitable depends on the amount of effort required of the predator to scan the habitat, the abundance of prey in the habitat, and the predator's ability to detect the prey, given that they are present. Certainly, this last consideration is affected by the predator's perceptual abilities. But it is also affected by the perceptual abilities of the prey, which, as we have seen, can be influenced by their pattern of social dispersion.

Treisman (1975a) also objected to the idea that a prey animal's ability to be detected depends on the magnitude of its minimum dimension. Based on human sensory data, he proposed instead that the ability of predators and prey to detect objects visually increases as the area of the object increases, the probability of detection being lowest when no object is present and reaching a maximum past a critical area.

Given that the predator scans the environment some optimal number of times, that its ability to detect prey depends on the size of the prey aggregation, and the additional assumption that concealment is the only benefit that the prey derive from living in a group (i.e., they can not escape once they detect a predator), Treisman showed that regardless of the number of prey in the habitat, the probability of detecting prey scattered randomly about the habitat is always greater than the probability of detecting prey living in a group. Because the prey cannot escape once they are detected, however, the per capita probability of being killed is greater for animals living in a group than for animals living alone. Thus, if the prey's only defense is concealment, natural selection favors individuals living alone.

Animals capable of escaping from a predator are involved, with the predator, in a contest for priority of detection. In such situations, Treisman showed that regardless of the size of the group, the collective ability to detect the predator always exceeds the increased conspicuousness of the group. Therefore, when prey can flee from a predator, natural selection favors animals living in groups.

But as Treisman indicated, these models are somewhat artificial. For example, they predict that there should be no limit to the size of the group. Even though they incorporate the perceptual abilities of both the predator and the prey, they fail to account for other activities that compete for the prey animal's time. Some of these activities, such as acquiring food, are just as important for survival as is the need to avoid predators. Thus, we should ask: Are animals in groups more likely to enhance their feeding success than solitary animals? And, if so, under what circumstances?

B. Resource Competition

Changes in social structure often accompany changes in the distribution and abundance of resources. For example, Brown (1964) accounted for the existence of territories by the principle that a resource is defended only when the energetic costs of defense are exceeded by the energetic gains associated with the exclusive use of that resource. Field evidence, such as that provided by Gill and Wolf (1975) on African sunbird energetics, supports the principle of economic defensibility. At normal nectar concentration, individual sunbirds defended feeding territories. But as the nectar quality improved, the cost of repelling invaders disproportionately increased. Eventually, the energetic costs of defense exceeded the energetic gains derived from the exclusive use of the flowers, and the birds abandoned their territories.

There is also evidence that economic considerations determine when groups should form. Zahavi (1971) altered the pattern of social organization of wintering wagtails by manipulating the dispersion of their food. When the food was distributed in small patches, the birds established territories, but when the food was presented in large clumps, which were unevenly distributed, the birds formed flocks.

More detailed field studies comparing the foraging success of individuals, in and out of flocks, also suggest that birds in groups enhance their foraging efficiency. Murton (1971a,b) showed that solitary wood pigeons pecked more slowly than did those in flocks, and Rubenstein and his co-workers (1977) showed that although all finches pecked at the same rate, those finches in flocks (both mixed- and single-species flocks) fed for longer uninterrupted feeding episodes than did solitary finches. Krebs (1974) also demonstrated that the amount of food eaten by individual great blue herons increased as flock size increased.

But why do individuals in groups have enhanced feeding success? Horn (1968) suggested that birds nesting in colonies minimize their daily foraging route. Based on a simple model, he concluded that in order to minimize the distance traveled while foraging, birds should disperse their nests about the habitat when food is evenly spaced but that birds should clump their nests when food is present in patches that are unevenly distributed in space and time. His field observation revealed that Brewer's blackbirds, which nest colonially, fed on prey that were indeed distributed in patches that appeared randomly throughout the habitat. But as Horn noted, because he did not study the feeding habits of birds that lived alone, he was unable to give conclusive support to his generalizations.

Others argue that individuals in groups enhance feeding success by acquiring certain types of information from neighbors. Ward and Zahavi (1973) examined the feeding patterns of many species of colonial nesting birds and concluded, on the basis of circumstantial evidence, that colonies act as "information centres." They claimed that birds that were unsuccessful at locating unevenly dispersed food resources could enhance their feeding success by following already successful individuals to the feeding areas.

It was Krebs (1974), however, studying a population of colonial nesting great blue herons, who provided convincing evidence that colonies can in fact serve as information centers. He showed that birds faced with an unpredictable food source left the colony on foraging trips synchronously and that nearest neighbors showed the most similarity with respect to time of departure and feeding destination.

For birds that aggregate only to feed, other types of information can be transferred among group members. Krebs and his co-workers (1972) have shown that chickadees in aviaries increased their foraging success by copying feeding actions of their flockmates. Field studies of the feeding behavior of finches also revealed the occurrence and beneficial effects of social learning (Rubenstein et al., 1977).

Other studies suggest that the physical presence of neighbors facilitates foraging. Often group cooperation is necessary to capture large prey. This appears to be the case for many mammal species (Kruuk, 1972; Schaller, 1972). At other times, group activity, because of high levels of disturbance, flushes prey from under cover. Many tropical flocks of birds forage in this manner (Willis, 1966; Morse, 1970).

Although these advantages often accrue to individuals living in groups, not all animals live in or feed in groups. In these instances, the costs of intensified competition most likely offset the potential gains, thus making group living, based solely on dietary consideration, uneconomical. Using a mathematical model based on a bird's ability to locate food, the abundance of food in the habitat, and the costs associated with aggression, Pulliam (1976) outlined some of the conditions under which animals should form feeding groups. He concluded that at low levels of food abundance the feeding rate of a dominant individual (one who can drive away a competitor) and a submissive individual would be higher than that of an individual feeding alone. Thus, under low levels of food abundance, natural selection would favor individuals foraging in a group. At higher food concentrations, however, Pulliam predicted a different outcome. At higher food levels, the dominant bird reaches its maximum feeding rate and begins to use its excess time to chase the subordinate bird. Because the subordinate is spending time fleeing from the dominant, its feeding rate is reduced. When the dominant bird's aggression reaches a level where the subordinate bird's feeding rate falls below that of a solitary bird, the subordinate leaves the group. Thus, based on Pulliam's model, animals are induced to form feeding groups only when food abundance is low.

Both empirical and theoretical studies indicate that the economics of foraging plays a major role in determining patterns of social organization. Different species, however, appear to respond to different aspects of the resource. For example, whereas decreased levels of food abundance seem to favor the formation of flocks in granivorous birds, increased food patchiness and unpredictability seem to favor flock formation in Brewer's blackbirds and wagtails. Thus, any real understanding of how feeding considerations affect an individual's decision to live either alone or in a group requires a knowledge of what the individual preys upon and the prey's overall relationship with the habitat.

V. PREDATION AND RESOURCE COMPETITION: THEIR COMBINED ACTION

From the preceding discussions, it appears that either the influence of predators or the competitively induced need to forage efficiently could alone provide individuals with advantages that under certain conditions could induce them to form groups. In fact, many biologists contend that one force or the other is solely responsible for the evolution of group living and that the other force is responsible for generating secondary adaptations. For example, Alexander (1974) has proposed that predation is solely responsible for the formation and maintenance of groups, whereas Ward and Zahavi (1973) have claimed that "the primary importance of predation in the evolution of information centres lies in its 'shaping' the assemblies (which are formed for the efficient exploitation of a patchy food supply) so that the resultant vulnerability to predation is minimized" (p. 532). Although this is not a trivial debate, it has polarized the study of sociality and has diverted attention away from the fact that animals living in groups must simultaneously contend with the conflicting needs of defense and nourishment. Certainly, variations with respect to the environment, the individuals, and the species affect the relative importance of each force. But any theory that claims to account for why animals living in groups remain in those groups must account for the combined effects of both selective forces.

A. Games against Nature

Any level of vigilance exhibited by a prey animal has a cost associated with it. Obviously, the more time a prey animal spends searching for a predator, the less time it can devote to other activities, such as feeding, grooming, or finding and securing mates. But in addition, Treisman (1975a) has shown that a prey animal can increase its probability of detecting a predator only by increasing its probability of mistakenly detecting a predator (i.e., by increasing its level of skittishness). Paradoxically, the more vigilant a prey, the more it will be in flight.

Because of these costs associated with vigilance, Treisman (1975b) has developed an economic model of social organization, which I call a *game against nature*. In this game against nature, the fitness of a prey animal depends not only on its ability to determine when a predator is present but also on its ability to determine when a predator is not present. Depending on the accuracy of the prey animal's perception, four possible situations can occur, each potentially providing a different fitness payoff ($V_{..}$). The following matrix depicts one possible payoff scheme.

Prey's response

		Detect	Not detect
Predator:	Present	$V_{PD} = -1$	$V_{PN} = -10$
	Absent	$V_{AD} = -1$	$V_{AN} = +1$

It reveals that if a prey animal detects a predator when a predator is present, it will interrupt whatever activity it is engaged in and flee the habitat, incurring a small cost. If it "detects" a predator when a predator is not present, then it also flees the habitat and incurs the same small cost. But if it fails to detect a predator when one is present, then it is killed, incurring a much larger cost. Conversely, if a prey animal correctly perceives that there is no predator in the habitat, it will continue whatever activity it is engaged in, thus receiving an increase in fitness.

These relative payoffs play an important role in determining a prey animal's overall expected fitness, but they are influenced by the probability of a predator's being present, the probability of a prey animal's detecting a predator, and the probability of a prey animal's incorrectly detecting a predator. Once an animal's expected fitness is determined, we can answer questions such as whether an animal living in a group receives a higher fitness than an animal living alone and, if so, for what environmental conditions, group sizes and levels of vigilance.

Treisman's game against nature not surprisingly shows that the expected fitness of prey animals living in a group increases as the probability of either the predator's being present or the prey's detecting the predator increases. But it also reveals that forming larger and larger groups does not automatically lead to increased expected gains. Instead, the model shows that there are group sizes that maximize an individual's expected gains. Optimal group sizes result from the fact that although the group's collective ability to detect predators increases as group size increases, the group's collective probability of false detection disproportionately increases as group size increases. Thus, as group size increases, the costs associated with large false detection levels eventually exceed the benefits associated with true detection levels, and the expected gains begin to decrease.

When the probability of predation changes, Treisman suggested that there are a number of ways that animals in groups can adjust their behavior to maximize their expected fitness. If the risk of predation increases, then individuals of species unable to adjust sensory capabilities can increase their fitness by living in larger groups. But individuals of species capable of adjusting sensory capabilities can also reduce the risk of predation by increasing their levels of vigilance and by living in smaller groups. Thus, it appears that similar environmental forces can affect behavior in different ways for different species.

Treisman addressed the issue of whether animals living in groups have higher expected gains than animals living alone by using a version of the game-against-nature model that includes the predator's detection capabilities. He found that prey animals in groups do not always have an advantage over their solitary counterparts. For most realistic parameter values of the model, it appears that when the number of prey animals in a habitat is low, animals do better when assembled into a group. But when the population of prey animals is large, the greatest protection from predators occurs when prey are distributed randomly about the habitat.

Treisman's game against nature has provided a powerful means of investigating how and under what conditions predation induces individuals to live in groups. It effectively shows how subtle variations in the accuracy of both the predator's and the prey's sensory capabilities can have dramatic impacts on social dynamics by altering an individual's chances of survival. In spite of these attributes, the model ignores some important biological considerations. Because he was concerned primarily with the influence of predation, Treisman deliberately ignored the direct influence of feeding considerations. Nevertheless, any theory of social organization should incorporate the influence of both forces. In addition, he assumed, probably for mathematical simplicity, that all individuals in a group receive the same expected fitness payoff. This assumption is certainly unrealistic because evolution occurs precisely because individuals compete and derive unequal benefits.

In the next section, I present a model, in the form of a game, that directly analyzes the conflict between the need for defense and the need for nourishment, without assuming that all individuals in a population receive the same expected payoff.

B. Social Games

If the benefits of group living are not shared equally among group members, then under certain conditions it might be advantageous for the individuals receiving the fewest benefits to abandon a group and live alone. Thus, we might ask: Could this type of behavior occur? And, if so, under what circumstances?

The theory of games provides a convenient means by which we can predict when animals should choose one strategy over another. The simplest social game would contain only two strategies: to live in a group or to live alone. In general, an animal adopts the strategy that produces the highest fitness. But because the fitness associated with each strategy depends not only on environmental conditions but also on individual attributes such as age, sex, size and past experience, the "best" strategy is often difficult to predict. In the following social games, we assume for convenience that both animals are of equal age, and we call the individual best at acquiring resources the *dominant* and the other individual the *subordinate*.

An individual's fitness during a time interval, x to $x + 1$, can be represented (as a first approximation) as:

$$V_x = p_x m_x$$

where p_x is the probability of living through the interval and m_x is the number of offspring produced during the interval. (A glossary of terms appears in the appendix.) Let the probabilities of a dominant and a subordinate individual living through the interval, while either living alone or in a group, be P_{xR}, p_{xr}, P_{xG}, and p_{xg}, respectively. In a similar fashion, let the number of offspring produced during the interval by a dominant or a subordinate individual, either living alone or in a group, be M_{xR}, m_{xr}, M_{xG}, and m_{xg}, respectively. Then the following matrix depicts the strategies of both the dominant and the subordinate animal, as well as the fitness payoffs that each would expect to receive.

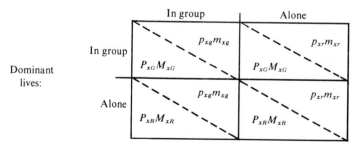

An individual maximizes its fitness only if it adopts an evolutionary stable strategy (ESS). According to Maynard Smith (1976, 1977), a strategy is evolutionarily stable only if a pair of individual strategies (S_D, S_s) exists so that it does not pay the dominant individual to diverge from its strategy S_D, when the subordinate adopts strategy S_s, and it does not pay the subordinate to diverge from strategy S_s as long as the dominant adopts strategy S_D. In the simple social game, there are four possible ESSs:

1. Dominant and subordinate live in a group: requires that $P_G M_G > P_R M_R$ (otherwise the dominant will live alone) and that $p_g m_g > p_r m_r$ (otherwise the subordinate will live alone).
2. Dominant in group and subordinate alone: requires that $P_G M_G > P_R M_R$ (otherwise the dominant will live alone) and $p_r m_r > p_g m_g$ (otherwise the subordinate will live in a group).
3. Dominant alone and subordinate in group: requires that $P_R M_R > P_G M_G$ (otherwise the dominant will live in a group) and that $p_g m_g > p_r m_r$ (otherwise the subordinate will live alone).
4. Dominant and subordinate live alone: requires that $P_R M_R > P_G M_G$ (otherwise the dominant will live in a group) and $p_r m_r > p_g m_g$ (otherwise the subordinate will live in a group). To determine under what conditions a potential ESS becomes an actual ESS, more precise descriptions of the p_x and m_x functions are required.

The functions that describe the probability of survival (p_x) are based on the idea that once an animal acquires its maintenance energy, the major factor affecting its probability of survival is its ability to detect a predator before the predator detects it. Thus, in general, the probability that an individual that lives alone will survive from age x to $x + 1$ is:

$$P_{xR} = p_{xr} = \alpha_x \cdot f_1[C_i(N), D_i(N)]$$

where α_x is the probability that an animal aged x will live to age $x + 1$, given that there are no predators in the habitat, and $f_1(C_i(N), D_i(N))$ is a discrete function describing the probability of an individual's not being killed by a predator during the interval x to $x + 1$. This function increases monotonically as the size of the population (N) increases and is influenced by the predator's ability to detect prey (C_i) and the prey's ability to detect a predator (D_i).

The probability that an individual living in a group will survive from age x to age $x + 1$ is:

$$P_{xG} = p_{xg} = \alpha_x \cdot f_2[C_G(N), D_G(N)]$$

where $f_2[C_G(N), D_G(N)]$ again is a function describing the probability of an animal's not being killed by a predator. It increases monotonically because both the probability of a predator's detecting a group (C_G) and the ability of prey in a group to detect a predator (D_G) increases as the size of a group increases.

The functions describing an animal's fecundity are based on the idea that the number of offspring that an animal can produce is related to the amount of energy that it can acquire and devote to reproduction. For an animal living alone, the number of offspring that it can produce during the interval x to $x + 1$ is:

$$M_{xR} = m_{xr} = \beta_x \cdot \gamma_x \cdot g_1 (r, N)$$

where β_x is a conversion factor that represents the ability of an animal aged x to convert calories into offspring, γ_x is the maximum amount of calories that an animal aged x can acquire for reproduction given that there are no competitors present, and $g_1 (r, N)$ represents the proportion of food actually acquired for reproduction as a result of an individual's competitive ability (r) and the number of competitors (N). This function decreases as either an individual's competitive ability decreases or as population size increases.

For an animal living in a group, the number of offspring that it can produce during the interval x to $x + 1$ is:

$$M_{xG} = m_{xg} = \beta_x \cdot \gamma_x \cdot g_1 (r, N) \cdot g_2(N)$$

where $g_2(N)$ represents the proportion of food eaten as the result of beneficial effects of neighbors. This function increases monotonically as group size increases.

Thus, in general, an animal aged x living in a group will have a fitness of

$$V_{xG} = V_{xg} = \alpha_x \cdot f_2 (C_G(N), D_G(N)) \cdot \beta_x \cdot \gamma_x \cdot g_1 (r, N) \cdot g_2(N)$$

whereas an equally old individual living alone will have a fitness of

$$V_{xR} = V_{xr} = \alpha_x \cdot f_1 (C_i(N), D_i (N)) \cdot \beta_x \cdot \gamma_x \cdot g_1 (r, N)$$

In order to predict which strategy an animal should choose, one must calculate fitness values for each strategy. Therefore, specific functions for f_1 ($\cdot$), f_2 ($\cdot$) g_1 ($\cdot$), and g_2 ($\cdot$) must be derived. The following g_1 ($\cdot$) and g_2 ($\cdot$) functions were chosen because they are simple and because the parameters governing their shape are sensitive to ecological consideration; functions based on physiological considerations would have been selected had enough data been available.

The maximum number of prey that are potentially supportable by a habitat is ultimately determined by the total amount of energy available in the habitat. If no competition occurred between members of the population, then each individual would be capable of acquiring the maximum amount of energy for reproduction (γ_x). As a result, only $N = T/\gamma_x$ individuals could coexist in the habitat. (T represents the total amount of available energy in the habitat.) Even if an inequitable distribution of resources occurred because of competition, it is assumed for the purposes of this model that the habitat can still support only N individuals. The excess energy is either lost through aggression or used by subordinate individuals to meet increased maintenance costs associated with increased levels of stress. Thus, with the size of the population fixed by external conditions, the negative effects of competition depend only on an animal's competitive ability (rank for animals in a group), the size of the population, and the level of competitive inequality that exists among the individuals in the population. One of the simplest functions incorporating these considerations is

$$g_1 = 1 - [(r - 1)/N]^a$$

where r stands for an individual's rank (competitive ability), N stands for the size of the population, and a is a coefficient that reflects the magnitude of competitive inequality in the population ($0 < a < \infty$). For example, when $a = 1$, the proportion of γ_x that an animal receives decreases linearly as its rank decreases (increase in r). When $a < 1$, the inequality in resource distribution is accentuated and the relationship becomes concave. Conversely, when $a > 1$, the inequality in resource distribution is diminished and the relationship becomes convex. An important property of the function is that for animals living in a group, even if animals are of equivalent rank, it takes on different values depending on the size of the group. For example, given two animals of equal rank, the one living in the smaller group will receive a greater proportion of γ_x than the one living in the larger group. Thus, regardless of competitive ability, this function indicates that animals do worse in large groups, where competition is more intense.

For animals living in groups, the presence of neighbors often enhances an individual's feeding success. The simplest increasing function that accounts for this effect is

$$g_2 = 1 + [b(N - 1)^c]$$

where b and c are parameters that govern the shape of this relationship. Values of b and c can range from ($0 < b$ or $c, < 1$); when b is small, increasing the number of neighbors (N) enhances feeding success very little. When c is also small, this effect is reduced even further, especially at larger group sizes.

The parameters a, b, and c can take on values that reflect a variety of environmental situations. For example, the value of parameter a could be increased to reflect increasing similarity in the competitive abilities of group members or an increasingly even distribution of food resources. In both cases, the higher value of parameter a would simulate a lower level of competitive inequality among the members of the group. Similarly, the values of parameters b and c could be increased to reflect increasingly concentrated or more patchily distributed food resources. In either case, higher values of the parameters would simulate greater feeding benefits to be derived from neighbors. This flexibility of the model's parameters provides a means of interpreting how various attributes of the resource under contention affect an individual's decision to live alone or in a group.

Because Treisman (1975a) provided realistic functions describing a prey's probability of being killed by a predator, I have modified them slightly so that they can serve as the $f_1(\cdot)$ and $f_2(\cdot)$ functions.

Treisman (1975a) proposed that the probability of a solitary prey animal's detecting a predator is $D_i = zA_i + P_f$, where z is a coefficient, A_i is the area of an object (e.g., a predator), and P_f is the probability of the prey's mistakenly detecting a predator when none is present. Similarly, he proposed that $C_i = zA_i + P_f$ is the probability of a predator's detecting a solitary prey. When N prey are present in the habitat, the probability of the predator's detecting at least one of the prey animals becomes

$$C_R = 1 - (1 - C_i)^N$$

which is one minus the probability of the predator's not detecting any prey. The probability of the predator's killing at least one prey animal is

$$K_{NR} = 1 - (1 - K_R)^N$$

where K_R is the per capita probability of a prey animal's being killed. In turn

$$K_R = \frac{(C_R/N)(1 - D_i)}{(C_R/N)(1 - D_i) + D_i}$$

where C_R/N is an individual's risk of being discovered. Thus, the probability of any individual's not being killed by a predator on any given search of the habitat is

$$f_1 = 1 - K_{NR}$$

If we can assume that a prey animal's risk of being killed on a particular search is indicative of its risk of being killed during the interval x to $x + 1$, then the probability that an animal that lives alone will survive from age x to age $x + 1$ is

$$P_{xR} = p_{xr} = \alpha_x \cdot (1 - K_{NR})$$

Animals in groups benefit from the scanning behavior of neighbors. As a result, the probability that an individual living in a group will detect a predator is

$$D_G = 1 - (1 - D_i)^{N \cdot a'}$$

where a' is a parameter that governs the actual number of individuals scanning for the predator (a' = min $(a, 1)$). The value of parameter a' depends on the value of parameter a because it is assumed that animals receiving few resources either make ineffective lookouts or disturb otherwise competent lookouts. Thus, when the inequalities in competitive payoffs are severe (low values of parameter a), the number of effective watchers in a group will be small (low values of parameter a').

Because a group presents a larger area to a predator, the probability of its detecting a group is C_G = z min $(NA_i, X) + P_f$, where X represents a threshold after which further increases in area no longer increase a predator's detection abilities. The probability that a predator will kill a prey animal living in a group during a scan is

$$K_G = \frac{C_G (1 - D_G)}{C_G (1 - D_G) + D_G}$$

Hence, the probability that a prey animal that lives in a group will not be killed is $f_2 = (1 - K_G)$. Its chances of living from age x to age $x + 1$ are

$$P_{xG} = P_{xg} = \alpha_x \cdot (1 - K_G)$$

For animals playing the social game based on the above functions, the fitness of an individual living in a group is

$$V_{xG} = V_{xg} = \alpha_x \cdot (1 - K_G) \cdot \beta_x \cdot \gamma_x \cdot (1 - [(r - 1)/N]^a) \cdot b(N)^c$$

whereas for an individual living alone it is

$$V_{xR} = V_{xr} = \alpha_x \cdot (1 - K_{NR}) \cdot \beta_x \cdot \gamma_x \cdot (1 - [(r - 1)/N]^a)$$

For the following simulations, let us assume that competition for resources is sometimes more intense among animals in groups than among solitary animals. This assumption is not unreasonable because animals in groups experience both exploitation and interference competition, whereas animals randomly dispersed (no territoriality) and separated by large distances are less likely to experience interference competition. Therefore, the values of parameter a for animals living in a group should be less than or equal to the values of a for prey distributed randomly throughout the habitat. Also, let us assume that animals living alone are less able to assess each other's competitive ability than animals living in a group. Therefore, as a first approximation, an animal leaving a group will do so with the "expectation" of deriving the benefits of the "average" solitary individual.

The results of simulations in which the parameters a, b, and c were varied are shown in Figs. 1–4. Each figure depicts fitness (V_x) as a function of population size. A solid isocline shows how the fitness of equally ranked individuals changes as the group size changes. The bottom of the hatched area shows how the fitness of an "average" ($r = 0.5\ N$) solitary individual changes as the number of animals living alone changes. The top of the hatched area similarly depicts the fitness changes of the "best" solitary competitor.

In all the simulations, fitness of group members varies according to rank. Not surprisingly, the magnitude of these fitness differences reflects the level of competitive inequality. When the intensity of competition is high (Fig. 1), the difference in the fitness of the dominant and most subordinate individual is larger than when the level of competitive inequality is low (Fig. 2).

Regardless of environmental conditions, the dominant individual always derives a higher fitness by being in a group of N individuals than by living alone as 1 of N randomly dispersed individuals. Thus, based on this specific form of the model, Strategy 3 (where the dominant lives alone and the subordinate lives in a group) and Strategy 4 (where both the dominant and subordinate live alone) of the social game cannot be evolutionarily stable.

The simulations reveal that the best strategy for a subordinate individual, given that the dominant lives in a group, does depend, however, on a number of environmental considerations. For example, when the visibility of the habitat is low, the level of competitive inequality among group members is high, the beneficial affects of neighbors is small (Fig. 1), and the habitat can support only 9 or fewer individuals, then the most subordinate individual in the group is always better off as 1 of 9 or fewer randomly dispersed individuals. As a result, it should leave the group. If the habitat can support more than 9 individuals, each individual by living in a group will receive a higher fitness than by living as one of 10 or more solitary individuals. But this comparison is somewhat misleading. Imagine that the habitat can support only 10 individuals; then, if the most subordinate individual were to leave the group and become the lone solitary individual, it could derive a large increase in its fitness. On the basis of this comparison, the most subordinate individual should leave the group. Should the 9th-ranking individual also leave the group? It should do so only if its fitness, as 1 of only 2 solitary individuals, will be higher than its fitness will be if it remains in a group of 9. Based on the assumption that an animal's percentile rank does not change when poorer competitors leave the group, we could sight along the 90th percentile isocline and determine what its fitness would be in a group of 9 individuals ($V_x = 0.24$) and then compare this value with the fitness of an "average" solitary individual ($V_x = 2.94$) belonging to a population consisting of 2 individuals. In this case, it appears that the 9th-ranking member should also leave the group. By the time the 6th member is faced with the choice of leaving the group, the payoffs have changed, so that it should leave the group if it could be 1 of only 4 randomly dispersed individuals. But if the 6th individual left the group, it would be 1 of 5 randomly dispersed individuals ($V_x = 0.52$), and under these circumstances, it should remain in the group, where its fitness is higher ($V_x = 0.76$). Thus, under the conditions of this simulation, for a habitat able to support 10 individuals, 6 would live in a group and 4 would live alone.

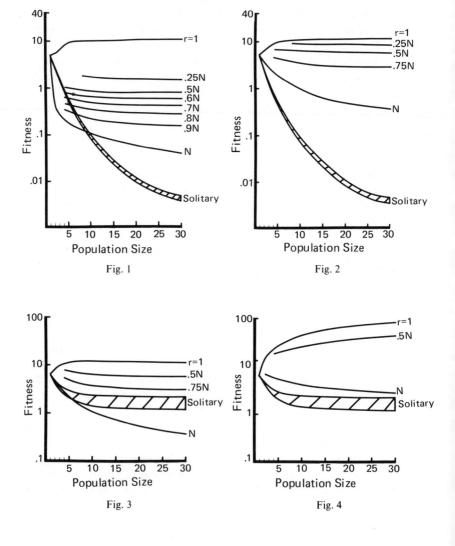

Fig. 1

Fig. 2

Fig. 3

Fig. 4

An interesting phenomenon occurs as the quality of the habitat increases and it is able to support more animals. For example, when the habitat can support 20 individuals, only the 5 lowest-ranking individuals would be induced to leave the group (Fig. 1). Thus, when food abundance increases in a habitat, a smaller percentage of the group will leave. Even though animals of equal percentile ranks get proportionately less of γ_x as the size of the group increases, the benefits of collective predator defense increase disproportionately as group size increases. Therefore, as the size of a group increases, its members find remaining in the group more profitable than living alone.

←——

Fig. 1. Relationship between the expected fitness of an individual and the size of the population, given that the visibility of the habitat is low ($C_i = D_i = 0.1$; $C_G = 0.8$), that the level of competitive inequality among members in a group is much more severe ($a = 0.1$) than among randomly dispersed individuals ($a = 2$), that the beneficial effects of neighbors are small ($b = c = 0.1$), and that $\alpha_x = 1$, $\beta_x = 0.1$, and $\gamma_x = 100$. The hatched area depicts the expected fitness of an animal aged x living alone as one of the N randomly dispersed individuals in the population; the bottom line reveals the expected fitness of the "average" solitary individual, whereas the top line reveals the expected fitness of the solitary individual that is the best competitor. The solid isoclines depict the expected fitness of an animal aged x living in a group. Each isocline connects the expected fitness of an animal of a given rank but living in groups of different sizes. The most dominant animal's fitness isocline is labeled $r = 1$, whereas the most subordinate is labeled N. The other isoclines correspond to animals in the 90th percentile ($0.9N$), the 80th percentile ($0.8N$), and so on. If we assume that the habitat can only support 10 individuals and all are originally arranged in a group, the open circles show that the expected fitness of the sixth ranking individual would be higher if it remained as the sixth ranked individual in a group of six than if it left the group and lived as one of five randomly dispersed individuals. See the text for a more complete description, but for this case, six individuals will live in a group and four will live randomly dispersed about the habitat.

Fig. 2. Relationship between the expected fitness of an individual and the size of the population, given that the visibility of the habitat is low ($C_i = D_i = 0.1$; $C_G = 0.8$), that the level of competitive inequality among members of a group is only slightly more severe ($a = 1$) than among solitary individuals ($a = 2$), that the beneficial effects of neighbors are small ($b = c = 0.1$), and that $\alpha_x = 1$, $\beta_x = 0.1$, and $\gamma_x = 100$. The hatched area depicts the expected fitness of animals living alone, and the solid isoclines depict the expected fitness of animals by rank living in a group.

Fig. 3. Relationship between the expected fitness of an individual and the size of the population, given that the visibility of the habitat is high ($C_i = D_i = 0.4$; $C_G = 0.8$), that animals living alone or in a group experience the same level of competitive inequality ($a = 1$), that the beneficial effects of neighbors are low ($b = c = 0.1$) and that $\alpha_x = 1$, $\beta_x = 0.1$, and $\gamma_x = 100$. The hatched area depicts the expected fitness of animals living alone, and the solid isoclines depict the expected fitness of animals by rank living in a group.

Fig. 4. Relationship between the expected fitness of an individual and the size of the population, given that the visibility of the habitat is high ($C_i = D_i = 0.4$; $C_G = 0.8$), that animals living alone or in a group experience the same level of competitive inequality ($a = 1$), that the beneficial effects of neighbors are high ($b = c = 0.7$) and that $\alpha_x = 1$, $\beta_x = 0.1$, and $\gamma_x = 100$. The hatched area depicts the expected fitness of animals living alone, and the solid isoclines depict the expected fitness of animals by ranking living in a group.

As the level of competitive inequality is reduced (Fig. 2), regardless of the carrying capacity of the habitat, a greater percentage of individuals will live in groups. For example, if the habitat can support 10 individuals, only the bottom 2 will be induced to leave the group. The 8th-ranking individual should not leave the group because it could raise its fitness by living alone only if it could be 1 of 2 randomly dispersed individuals. But if it leaves the group, it will be 1 of 3 randomly dispersed individuals. Thus, it appears that as differences in ability among individuals decrease or as evenness of the resource distribution increases, larger groups will form.

It is interesting to note that as the parameter a increases in value, the fitness of all percentile levels increases. In part, this increase results from the fact that the percentage of effective watches increases (a' increases as well), but because the percentage of changes in fitness is inversely correlated with rank, the effects of competition should not be ignored.

In more open habitats, where both the predator and the prey are more likely to detect each other (see Figs. 3 and 4), some interesting changes in behavior occur.

If an open habitat can support only 5 or fewer individuals and the level of competitive inequality among individuals is the same for individuals living either alone or in a group, then each individual will do better living in a group as opposed to living as 1 of 5 or fewer solitary animals (Fig. 3). As before, however, this comparison is somewhat misleading. For example, if a habitat can support 10 individuals (Fig. 3), only the bottom 3 individuals will be induced to leave the group. By comparison, in the previous example (Fig. 2), in which parameter a also equaled 1, only 2 members of a group of 10 were induced to leave the group. Thus, it appears that as habitat visibility improves, the size of the groups residing there will be smaller because in open habitats, the advantages of many eyes' assisting in detecting a predator is diminished. Thus, as long as competition among individuals in a group is at least as severe as that among solitary individuals, the relative risks of living alone decrease in more open habitats.

When the presence of neighbors greatly enhances an individual's feeding success, then there are habitat carrying capacities at which it always pays for all animals, regardless of rank, to live in a group. For example, under the conditions of the simulation depicted in Fig. 4, if the habitat can support only 2 individuals (actually almost 3 individuals), each individual will derive its highest fitness by living in a group. For a habitat capable of supporting 10 individuals, only the most subordinate individual will derive a higher fitness by living alone. Thus, it appears that habitats in which the beneficial affects of neighbors are large, even if the benefits associated with collective predator detection are small, will be populated by large groups. (This effect would be attenuated somewhat if the level of competitive inequality were increased.)

Thus, for any given set of conditions, either Strategy 1 (both dominant and subordinate live in a group) or Strategy 2 (the dominant lives in a group and the subordinate lives alone) will be evolutionarily stable. But which of these two strategies an animal chooses will depend on its rank in the group. As external conditions change, or as the animal's competitive ability changes, it is likely that its choice of an evolutionary stable strategy will also change. The more dominant an individual, the more likely it is to live in a group, regardless of environmental conditions. The more subordinate an individual, however, the more likely it is to remain in a group when the visibility of the habitat is low, when the level of competitive inequality among group members is low, when the carrying capacity of the habitat is high, and when the presence of neighbors greatly enhances its feeding success.

VI. DISCUSSION AND CONCLUSIONS

We have seen that ethologists have made great strides toward formulating a predictive theory of social organization. Speculative explanations as to how the forces of predation and foraging considerations operate to influence the formation of a social system have been replaced by mathematical formulations. These formulations, which are based on biological principles, have generated many insights into why and under what environmental conditions groups could form. These formulations have also provided insights into what attributes best adapt individuals for particular types of social systems.

Whereas mathematical models were originally used to investigate the effects of predation or the effects of foraging demands, they are now beginning to be used to study the combined action of predation and foraging. In the social game model, general fitness functions, based on aspects of predatory and competitive behavior, are used to predict when and under what circumstances an animal should either live alone or live in a group. The importance of rank (competitive ability) is clearly demonstrated: under certain conditions subordinate individuals could acquire a higher fitness by leaving the group. Because many factors—such as age, size, sex, and past experience—influence competitive ability, this model provides a means of examining many dynamics of social behavior. For example, because competitive ability changes with age, the model predicts that animals of certain ages should temporarily abandon living in groups. In fact, additional insights into the importance of age effects can be derived if one simply allows the age-specific parameters, α_x, β_x, and γ_x, to vary.

Although this model investigates some new facets of the social organization problem, it leaves many untouched. For example, it makes no provision for the possibility that subordinate animals that leave the group might derive a higher fitness by forming a new group rather than by remaining totally solitary. This omission could be rectified if one increases the number of strategies to account for a variety of dispersion patterns and incorporates the effects of intergroup competition into the general fitness functions.

More importantly, however, this model excludes considerations of inclusive fitness (Hamilton, 1964) from the general fitness functions. Certainly, groups consisting of closely related individuals display a much lower level of competitive inequality. As a result, subordinate individuals would be less likely to leave such a group. But other, more subtle complications of inclusive fitness could arise. Imagine two daughters' being induced by fitness considerations to leave the group. It is possible that the mother might follow them, and by joining them in a triad, she might lower their risk of predation and thus increase her inclusive fitness. And, in fact, group size and inclusive fitness could create even greater complications (cf. Charnov, 1977).

The model also ignores some potentially important ecological considerations. For example, the effects of interspecific competition are normally considered less intense than the effects of intraspecific competition and, as a result, have been ignored in investigations of patterns of social organization. The presence of a competitor species (Species B), however, could indirectly have two major effects on the species under consideration (Species A): (1) the presence of competitor species B could lower the effective resource level of the habitat, and (2) it could drastically alter the predation pressures on Species A.

The change in predation pressure that could result from the presence of a competitor Species B might occur in a complex fashion. If the competitor Species B is abundant and is being disproportionately consumed by the predator, then Species A would experience a low level of predation intensity. But the fitness payoffs to Species A associated with this level of predation could change dramatically; after reducing the abundance of the competitor Species B, the predator might disproportionately concentrate its attack on Species A. Then, in a while, after decimating Species A, it might again switch its feeding preference to competitor Species B, or it might even leave the habitat completely. In either event, the intensity of predation suffered by Species A would again be low. The importance of interspecific competition, therefore, depends on the frequency and the predictability of the predator's switching behavior. If many such oscillations in its behavior occur during the time interval during which the fitness of an individual of Species A is being estimated, then the model would be too simplistic and would make unrealistic predictions concerning an individual's decision to live alone or in a group. The extent to which this phenomenon is important in influencing patterns of social organization is yet to be determined.

I dwell on this particular consideration only to stress that although great strides have been made toward the formation of a predictive theory of social organization, much more work is needed. In the past, ethologists have ignored subtle ecological influences such as this one. But just as interest in sociality continues to grow, the scope of the investigation must also broaden.

VII. ACKNOWLEDGMENTS

Many people have influenced the development of these ideas. I would like to thank B. Hazlett and R. D. Alexander for introducing me to the evolutionary study of animal behavior. Because these ideas grew out of my doctoral research on the mechanisms of competition, I would like to thank two of my committee members, Drs. P. H. Klopfer and H. M. Wilbur, for their continuing support, comments, and discussions. Thanks also go to A. Łomniki, who encouraged me to describe mathematically the effects of unequal competition; to L. Fairchild, J. Travis, and R. Wallace, who provided valuable comments on an earlier draft of the manuscript; and to C. Dewey, who helped prepare the final version of the manuscript. During the period that this research was carried out, I was supported as a National Science Foundation Graduate Fellow.

VIII. APPENDIX—GLOSSARY OF TERMS

A_i = area of an object

a = parameter reflecting level of competitive inequality

a' = parameter reflecting the proportion of competent predator lookouts

b = parameter reflecting the level of feeding enhancement

C_i = probability of a predator's detecting solitary prey

C_G = probability of a predator's detecting a group of prey

c = parameter reflecting the level of feeding enhancement

D_i = probability of solitary prey's detecting a predator

D_G = probability of a group of prey's detecting a predator

K_G = probability of predator's killing a prey living in a group

K_{NR} = probability of predator's killing at least 1 randomly scattered prey

K_R = per capita probability of a randomly scattered prey's being killed

min (x, y) = y or x, whichever is least

m_x = number of offspring produced during the time interval x to $x + 1$. (M_x = the fecundity of dominants; m_x = the fecundity of subordinates; M_{xG} or m_{xg} = the fecundity of dominant or subordinate in a group; M_{xR} or m_{xr} = the fecundity of solitary animals)

N = population size

P_f = probability of prey's mistakenly detecting a predator or of predator's mistakenly detecting a prey

P_x = probability of surviving from age x to age $x + 1$ (P_x = probability of dominant's surviving; p_x = probability of subordinate's surviving; P_{xG} or p_{xg} = probability of dominant or subordinate's surviving in a group; P_{xR} or p_{xr} = probability of dominant or subordinate's surviving as solitary animals)

r = competitive ability or rank

T = total amount of energy available in habitat

$V_{..}$ = fitness

V_x = fitness of an animal during interval x to $x + 1$

X_i = threshold past which increases in area no longer increase ability to detect an object

x = an animal's age

z = a coefficient

α_x = probability that an animal aged x will live to age $x + 1$ given that there are no predators in the habitat

β_x = conversion factor representing the ability of an animal aged x to convert calories into offspring

γ_x = maximum amount of calories that an animal aged x can acquire for reproduction, given that no competitors are present

IX. REFERENCES

Alexander, R. D. (1974). The evolution of social behavior. *Annu. Rev. Ecol. Syst.* 5:325–383.

Brown, J. L. (1964). The evolution of diversity in avian territorial systems. *Wilson Bull.* 76:160–169.

Charnov, E. L. (1977). An elementary treatment of the genetical theory of kin-selection. *J. Theor. Biol.* 66:541–550.

Emlen, S. T. (1975). Adaptive significance of synchronized breeding in a colonial bird: A new hypothesis. *Science* 188:1029–1031.

Gill, E. B., and Wolf, L. L. (1975). Economics of feeding territoriality in the golden-winged sunbird. *Ecology* **36**:333-345.

Hamilton, W. D. (1964). The genetical evolution of social behavior, I and II. *J. Theor. Biol.* **7**:1-52.

Hamilton, W. D. (1971). Geometry for the selfish herd. *J. Theor. Biol.* **31**:295-311.

Hoogland, J. L., and Sherman, P. W. (1976). Advantages and disadvantages of bank swallow (*Riparia Riparia*) coloniality. *Ecol. Monogr.* **46**:33-58.

Horn, H. S. (1968). The adaptive significance of colonial nesting in the Brewer's blackbird (*Euphagus cyanocephalus*). *Ecology* **49**:682-694.

Krebs, J. R. (1974). Colonial nesting and social feeding as strategies for exploiting food resources in the great blue heron (*Ardea herodias*). *Behaviour* **51**:99-134.

Krebs, J. R., MacRoberts, M. H., and Cullen, J. M. (1972). Flocking and feeding in the great *Parus major*—An experimental study. *Ibis* **114**:507-530.

Kruuk, H. (1972). *The Spotted Hyena*. Chicago University Press, Chicago.

Maynard Smith, J. (1976). Evolution and the theory of games. *Am. Sci.* **64**:41-45.

Maynard Smith, J. (1977). Parental investment: A prospective analysis. *Anim. Behav.* **25**:1-9.

McFarland, D. J. (1976). Form and function in the temporal organisation of behaviour. In Bateson, P. P. G., and Hinde, R. A. (Eds.), *Growing Points in Ethology*. Cambridge University Press, Cambridge, England, pp. 55-93.

Morse, D. H. (1970). Ecological aspects of some mixed-species foraging flocks of birds. *Ecol. Monogr.* **40**:117-168.

Murton, R. K. (1971a). The significance of a specific search image in the feeding behavior of the wood pigeon. *Behaviour* **39**:10-42.

Murton, R. K. (1971b). Why do some birds feed in flocks? *Ibis* **113**:534-536.

Pulliam, H. R. (1973). On the advantages of flocking. *J. Theor. Biol.* **38**:419-422.

Pulliam, H. R. (1976). The principle of optimal behavior and the theory of communities. *Perspect. Ethol.* **3**:311-332.

Rubenstein, D. I., Barnett, R. J., Ridgely, R. S., and Klopfer, P. H. (1977). Adaptive advantages of mixed-species feeding flocks among seed-eating finches in Costa Rica. *Ibis* **119**:10-21.

Schaller, G. (1972). *The Serengeti Lion*. Chicago University Press, Chicago.

Tinbergen, N., Broekhuysen, G. J., Feekes, F., Houghton, J. C. W., Kruuk, H., and Szulc, E. (1962). Egg shell removal by the black-headed gull, *Larus ridibundus* L: A behavior component of camouflage. *Behaviour* **19**:74-118.

Treisman, M. (1975a). Predation and the evolution of gregariousness. I: Models for concealment and evasion. *Anim. Behav.* **23**:779-800.

Treisman, M. (1975b). Predation and the evolution of gregariousness. II: An economic model for predator-prey interaction. *Anim. Behav.* **23**:801-825.

Vine, I. (1971). Risk of visual detection and pursuit by a predator and the selective advantage of flocking behaviour. *J. Theor. Biol.* **30**:405-422.

Ward, P., and Zahavi, A. (1973). The importance of certain assemblages of birds as "information-centres" for food-finding. *Ibis* **115**:517-534.

Willis, E. O. (1966). Interspecific competition and the foraging behavior of plain-brown wood creepers. *Ecology* **47**:667-674.

Zahavi, A. (1971). The social behaviour of white wagtail *Motacilla alba alba* wintering in Israel. *Ibis* **113**:203-211.

Chapter 10

IS HISTORY A CONSEQUENCE OF EVOLUTION?

L. B. Slobodkin

Ecology and Evolution Program
State University of New York, Stony Brook
and
Smithsonian Institution, Washington, D.C.

I. ABSTRACT

The recurrent idea that human history is a special case of the actions of evolutionary law and is therefore in some sense predictable from evolutionary law is, itself, of major historical importance.

This idea is shown to be invalid since human individual behavior is remarkably free of evolutionary constraints. In fact, humans have the biological property that much of the motivation for their behavior is related to their own socially derived image of themselves, which minimizes the significance of natural selection for or against specific human behaviors. The capacity to develop such an image is not widespread among animals but is shared by man and certain apes. The fact that evolutionary law is not coercive on human history has moral and political implications.

II. INTRODUCTION

Social behavior in animals is generally treated by evolutionary theorists as the evolutionary resultant of genetic and selective properties, as is any other behavioral or morphological feature. Learning is thought of as a mechanism for modification of behavior within limits set by the genetic capacity, so that we can consider the behavior of certain animals to be more poorly canalized (in the sense of Waddington, 1957) than that of others, and

233

the greater the significance of learning, the weaker the canalization of behavior. There is a persistent feeling that the range of observed behavior is kept within bounds by natural selection and that any observed behavior is almost certainly adaptive. If it doesn't appear adaptive at first glance, this is taken as an indicator of our lack of understanding of the total situation.

In this framework, the history of *Homo sapiens* is considered as being equivalent in kind to that of any other species in the sense that the behavior of humans is under the broad control of natural selection with implied genetic distinctions in fitness. This attitude has certain political implications, as does any attitude toward the deep structure of human history.

I will develop the position that human history differs profoundly from the history of other organisms and differs to such an extent that the usual modes of evolutionary thought, in which the actions of organisms are "judged" in terms of their selective value, are simply not applicable.

This argument will be based on biological properties of humans and some of the higher apes. In particular, I will claim that the biological capacity for the development of an environmentally determined self-image makes it impossible for human history, both public and private, to be predicted from considerations of evolutionary theory in any practically significant way.

It must be emphasized from the outset that I am not building my argument on either metaphysical or political grounds. For reasons that will be apparent from the text, I cannot unequivocally assert that I am not in some sense politically motivated in this presentation. I can, however, demonstrate that my intellectual concern derives from purely scientific consideration of the mechanisms of evolutionary strategy (Slobodkin, 1964, 1968; Slobodkin and Rapoport, 1974) and the difficulty of applying a theory of evolutionary strategy to humans (Slobodkin, 1977).

In comparing the idea of causality in history and evolution, I will show that in both cases, causality consists of defining a class of possible outcomes from a series of events rather than defining either a unique event or even a stochastic array of events.

I will then consider and reject the hypotheses that there are trends in evolutionary process and that since history can be thought of as a sequence of evolutionary events, we may be able to infer trends in history from trends in evolution.

I am concerned with the distinction between history as written by Braudel (1972) on the one hand and by Spengler (1934) on the other. In Braudel's text, there is a recital of historical events placed in an ecological, economic, and social context. There are certain generalities drawn, for example, that occupation of flat plains involves a stronger central government and stronger control of behavior than the occupation of mountains.

Braudel did not, however, infer any relation between inevitable sequences of historical events, nor did he consider that historical events are bound in any way by the biological evolutionary properties of men. By contrast, there is an enormously long intellectual history of the idea that human history is an inevitable consequence of either a supernatural or a natural destiny. Jan Kott (personal communication 1975) suggested that one of the primary intellectual contributions of Aeschylus and Aristotle was to deny the concept that human history is an incidental consequence of the destined history of the gods. That is, he believes that the pre-Aeschylan notion was, to a large extent, that the gods are bound in their own destiny, that men are in principle free, but that because the gods keep interfering in the affairs of men, the freedom is to a large degree illusory. Dr. Kott has also suggested that with the advent of Christianity as a major intellectual force, God becomes free of destiny and the inevitability of human history is put in abeyance. The classical concept of the Golden Age and the sequence of ages since the Golden Age was certainly remembered. This concept of a historical sequence, involving ages of man and a kind of inevitability, was regenerated as a result of the discovery of the New World, where the people of Europe saw the Golden Age of primitive man through the mist of distance.

With the discoveries of the 18th century and with the development of the theory of evolution, the ancient and recurrent analogy of society with either an organism or a sequence of organisms (a phyletic line) is made in a rather strong form. The later resurgence of the idea of historical inevitability hinges on the assumption that biology asserts that there is a kind of inevitability in the evolutionary process itself.

While I am not technically competent to discuss the nature of inevitability in history, I am competent to discuss inevitability in evolution. I will show that no matter how much one would like to do it, there is no way of inferring from the inevitability of biological evolutionary events the inevitability of historical events, since there are no inevitable trends in evolution, other than an ongoing, unchanging tendency toward conservative behavior (in an evolutionary, not a political sense).

In this assertion, I am confounding two separate notions. One is that a decision will inevitably fall within a historical or evolutionary trend since the decision maker is responding to the fact that he is gripped in this law whether he is aware of it or not (in the sense of Trivers, 1971, Wilson, 1975, and Alexander, 1974, 1975). The other notion is that the decision maker, in making his decision, does so with conscious cognizance of some trend that he believes to exist and that his decision is designed either to aid or to combat. The confounding is deliberate since I will show that there is no reason, based on evolutionary theory, to consider either the course of human history or the decisions made by individual decision-makers to be limited by evolutionary events.

The class of possible evolutionary futures is in one sense highly restricted by the properties of the evolutionary present and the conservatism of evolution. That is, the probability of turtles developing wings or penguins developing gills is very small. Not only is there a tendency not to do anything, but if anything does occur, it tends to occur in the smallest steps possible.

Sahlins (1976) has recently denied this kind of assertion. He quoted a statement by Wilson (1975), "If a female salmon laid only one or two eggs, the reproductive effort, consisting principally of the long swim upstream, would be very high. To lay hundreds more eggs entails only a small amount of additional reproductive effort" (1975, p. 97), which Sahlins criticized as follows:

> If selection will go so far as to atrophy the digestive tract in favor of a single reproductive explosion that also kills the organism, why should it not as easily effect structural changes that will allow the salmon to spawn twice or more to the same fitness effect, as for instance sturgeons do? (cf. ibid., p. 95). The problem is that this course or some other was precluded not by a natural selection but by an analytic one. The salmon was taken as an *a priori* limited being with one possible solution to the evolutionary problem of resource allocation to fitness, by a premise not motivated in the nature of evolution itself. The salmon is going to have only one chance to lay eggs, and that at very considerable cost. Once this set of conditions is taken as given, all other evolutionary possibilities to the same net fitness effect may be conveniently ignored. Since they are ignored, selection enters into the explanation as the mode of achieving an outcome intrinsic to the salmon. And the salmon's self-determined project of maximization becomes the logic of adaptive change. In other words, by the nature of the argument, the roles of the organism and natural selection in traditional evolutionary theory are perfectly reversed: The organism sets the orientation of change, while selection is assigned the function of providing the necessary materials. (p. 83)

Sahlins stated that Wilson had committed the "Fallacy of an *a priori* fitness course."

As a matter of fact, there is general agreement among evolutionary biologists that organisms are *a priori* limited in their possible evolutionary futures by their present state and by the fact that smaller evolutionary steps are more likely to be of value than larger ones. Similar statements have appeared for many years. For example, in 1961, I wrote:

> It is quite obvious that the effect of selection on [reproduction] will vary widely from animal to animal. If an animal has a very large litter size, as, for example, an oyster, it might be expected that natural selection would be more likely to increase litter size than to increase longevity of each female; the addition of one egg to the several million already present would have the same effect on [reproduction] as an infinite female reproductive life. An animal like man, which normally has only one young in a litter, faces severe physiological problems if the litter size is increased. Selection for increasing [reproduction] will be expected to result in an early onset of reproductive activity and a long reproductive life in species that have very small litters. (p. 54)

The effect on reproduction is just a special case of the classical argument presented by Sir Ronald Fisher (1930) favoring the position that evolution proceeds in smaller rather than larger steps, other things being equal. Fisher said:

> An organism is regarded as adapted to a particular situation, or to the totality of situations which constitute its environment, only in so far as we can imagine an assemblage of slightly different situations, or environments, to which the animal would on the whole be less well adapted, and equally only in so far as we can imagine an assemblage of slightly different organic forms, which would be less well adapted to that environment. . . .
>
> The statistical requirements of the situation, in which one thing is made to conform to another in a large number of different respects, may be illustrated geometrically. The degree of conformity may be represented by the closeness with which a point A approaches a fixed point O. In space of three dimensions we can only represent conformity in three different respects, but even with only these the general character of the situation may be represented. The possible positions representing adaptations superior to that represented by A will be enclosed by a sphere passing through A and centered at O. If A is shifted through a fixed distance, r, in any direction its translation will improve the adaptation if it is carried to a point within this sphere, but will impair it if the new position is outside. If r is very small it may be perceived that the chances of these two events are approximately equal, and the chance of an improvement tends to the limit ½ as r tends to zero; but if r is as great as the diameter of the sphere or greater, there is no longer any chance whatever of improvement, for all points within the sphere are less than this distance from A. For any value of r between these limits the actual probability of improvement is
>
> $$\tfrac{1}{2} \left(1 - \tfrac{r}{d}\right),$$
>
> where d is the diameter of the sphere.
>
> The chance of improvement thus decreases steadily from its limiting value ½ when r is zero, to zero when r equals d. Since A in our representation may signify either the organism or its environment, we should conclude that a change on either side has, when this change is extremely minute, an almost equal chance of effecting improvement or the reverse; while for greater changes the chance of improvement diminishes progressively, becoming zero, or at least negligible, for changes of a sufficiently pronounced character. . . .
>
> The conformity of these statistical requirements with common experience will be perceived by comparison with the mechanical adaptation of an instrument, such as the microscope, when adjusted for distinct vision. If we imagine a derangement of the system by moving a little each of the lenses, either longitudinally or transversely, or by twisting through an angle, by altering the refractive index and transparency of the different components, or the curvature, or the polish of the interfaces, it is sufficiently obvious that any large derangement will have a very small probability of improving the adjustment, while in the case of alterations much less than the smallest of those intentionally effected by the maker of the operator, the chance of improvement should be almost exactly half. (p. 38–40)

Fisher's argument refers to genetic change, but we will see that a similar favoring of small steps over larger ones also occurs on the phenotypic level.

To say that evolutionary steps tend to be small says nothing at all about which small steps actually occur, nor does it assert that a sequence of such steps will be constrained to occur in some specific direction.

We will indicate the biological peculiarities of man that are likely to influence historical events. We will deal with the curious fact that humans (and at least their close primate relatives) have a biological need to have a self-image, a concept of themselves, that they will act to preserve. We will also indicate that there is no clear evidence at all to support the assertion that one self-image is in some sense "more natural" than any other. That is, humans clearly have a self-image in the sense that librarians clearly have books. The content of the books is not ascertainable from their existence in the library. The content of self-image seems to be environmentally produced. If there are neurobiological limits to this process, they seem enormously broad.

An individual's self-image, once it has developed ontogenetically, becomes a powerful causal factor in determining what he or she does next and considerably broadens the class of possible future events that will result from any given social, historical, or environmental perturbation. We will already have shown that evolution occurs only in response to perturbation and has no momentum of its own.

Since men respond to their self-image as well as to their environment, it is obvious that a new level of looseness will have entered the relation between the magnitude of a response to an environmental change and the magnitude of that change itself. At the same time, it will be apparent that most of the responses do not have evolutionary meaning in the sense of changing genetic structure, although they do have historical meaning in the sense of being one of a class of possible sequences of events that might relate to biologically identical individuals.

Finally, we will demonstrate that while the self-image–forming capacity of man and the higher primates opens up historical possibilities and magnitudes of response that would not otherwise be available, man is not yet immune to elementary biological evolutionary changes of the simplest kind. How to respond to them becomes a question whose answer lies, of necessity, in the imagination of man and poses more problems than it solves.

We will then conclude from these arguments that the theory of the evolutionist can be of help to the historian only in freeing him of the need of looking for biological analogues. Decisions as to morality, as to what is or is not desirable about a political system, must be made by men without the help of evolution's telling them that there is a best way. However, one of the factors that must be considered in their decisions is the evolutionary effect of these decisions. Human decisions mold the evolutionary process. But human decision-makers cannot rely on evolutionary events or evolutionary theories to provide them with rectitude, safety, or sanctity.

On a completely different level, however, it will be apparent that beliefs about the evolutionary process are historically significant in the sense that what men believe about the "laws" that govern their history is a force in history. That is, beliefs about the biological nature of man strongly influence practical behavior quite aside from the specific problems of evolutionary theory itself. This fact may be seen in a variety of contexts. Perhaps one of the simplest is in the history of anti-Semitism. Putting it in an oversimplified form, the toleration, if not particular affection, of the Roman government for the Jews was based on the assumption that Judaism was another national religion, and in general, it was part of colonial policy not to interfere with local religious beliefs. By the later Middle Ages, Judaism was no longer a local religion of any sort, and the attitude of European Christianity toward the Jews was that they were persons of perverse opinion, either through ignorance or through a deliberate stubbornness of the will. When the Crusaders burned the Jewish towns of the Rhineland, the choice of conversion was available to the Jews because of the understanding that the nature of man was largely dictated by his own will. Several hundred years later, the simple statement of belief was no longer sufficient. That is, the theory had developed that men could have impurities in their explicitly stated beliefs and that these required a somewhat deeper analysis than simply asking the man. The original relation of the Spanish Inquisition to the Maranos and Moriscos was an attempt to get at the deep nature of belief in an apparently overtly believing Christian (Roth, 1964). The concern in 16th-century Spain with the purity of blood (satirized in Cervantes's *El retablo de las maravillas*) was the beginning of the modern assertion that men differ in their deep biological nature. The 19th-century development of evolutionary and genetic theory provided a rationalization for the belief in deep biological differences, so that the anthropological notions of Grant (1916), among others, strongly emphasized the supposed deep biological properties and provided the rationale for 20th-century anti-Semitism, which has really very little to do with the overt beliefs of its victims.

The Nazis did not provide the freedom to recant that was available to the victims of both the Crusaders and the Inquisition. The notion of survival of the fittest, derived from evolution, provided the rationale for the elimination of the unfit, where the definitions of fit and unfit were of the most curious and arbitrary kind.

It should be clear that the Nazis were not the only persons to see themselves as bearing the banner of evolution or as having the responsibility for the evolutionary history of mankind in their hands. Kipling's notion of the white man's burden represents a rather widely held rationalization for

British colonial policy of the 19th century. The fact that Karl Marx is reported to have asked Darwin for permission to dedicate *Das Kapital* to him (Barzun, 1958) is of interest, as is the recent study by Kamin (1974) of the role of evolutionary thought in the establishment of immigration policies, sterilization programs, and "eugenic" laws in the United States. There is an appeal to the idea that man is in the grip of evolutionary law and that politically, socially, and historically significant action must be taken to regulate the evolution of man.

We will be able only to put a little flesh on the bare bones that we have listed. There will not be time to give full documentation.

III. CAUSALITY IN EVOLUTION AND HISTORY

When one looks at the narrative of biological evolution or of the events of history, the apparent sequence of events is, to a degree, dependent on how the events are classified. Using a biological example, if we look at the class of animals that swim, we find that the ancestors of some members of that class consist only of animals whose ancestors have always been able to swim (fishes, protozoans, some crustaceans). In addition, we find other animals whose early ancestors could swim and whose less remote ancestors could not, but whose most immediate ancestor returned to swimming again. The tortoises that returned from the land swim by an adaptation of a tortoise walking gait. The whales and their relatives swim by something that is closer to wagging their tails than anything else, and it is rather similar to what the fish do. The ancestors of the whales were apparently carnivorous, terrestrial animals. One group of swimming birds, the penguins, adapted a flying motion to aquatic purposes, while the ducklike birds swim with their feet rather than with their wings.

Swimming, as a class of behaviors, arose from many different circumstances and different preadaptations. Therefore, the problems of adapting to water, with their array of possible solutions, may be viewed as a convergence of causal pathways to the development of swimming. We are talking of a many–one relationship in time. However, if I could make up for myself a series of words, one to represent what whales do and one to represent what penguins do and so on (and I am quite certain that just as Arabic has many words for camel and the Eskimos have many words for snow, somewhere a nation of frogmen will develop many words for swimming), I would then restore the appearance of one-to-one relationship or even a one-to-many relationship, and this relationship would be contingent on how carefully I classify my original animals. In short, while evolutionary events do not transpire at random, the definition of the kind of causality (that is, whether it is many–one, one–many, or one–one) becomes contingent on the classification of the events being considered. Is a democracy a democracy or are there 35 different kinds of democracies? Is a war a war or are there 70–100 different kinds of wars (see Vayda, 1976)? How one makes the decision determines what one gets in the way of a causal relation.

A way of proceeding as if the problem did not exist is what has been called *cybernetic causality*. The general idea is that a particular system, when acted on by an external stress or responding to the readjustments within its own structure, has a large class of impossible future states and another class of more-or-less equivalently possible states.

To put it another way, the evolutionary process forbids certain futures to evolving units but does not uniquely determine which of the nonforbidden states will actually arise. There are also certain historical transitions that will probably not occur, so that the probability of the Kalahari Bushmen's taking charge of West Africa is so small as to be set a zero, but the extinction (either cultural or physical) of the Kalahari Bushmen is within the class of possible events.

Note, by contrast, that the taxonomic problem in physics and chemistry is severely reduced. The chemical elements are known, and to some degree at least, the primary particles are known.

Should it prove to be the case that there are natural taxonomic units in historical events, then the causality issue can be solved in a more civilized fashion.

So far, we have loosened the notion of strict causality as applied to evolutionary events and (in almost exactly the same sense) to historical events. A qualification is required before the similarity receives too much attention.

In evolution, the fundamental laws of gene-frequency change (given certain kinds of natural selection and certain kinds of distribution of populations) are, in fact, amenable to detailed analysis. What causes more difficulty is what the meaning of those gene changes will be in terms of such things as swimming or hiding, running or smelling bad, becoming intelligent or becoming stupid, and so forth. That is, it is the functional consequences of a gene-frequency change that have a kind of difficult causality about them, analogous to historical causality (Lewontin, 1974). It may, perhaps, be the case that the biological properties of individual men have the same relation to the historical events in which those men participate as the biological properties of individual genes have to the evolutionary events in which those genes participate. That possibility requires further analysis elsewhere.

It might be considered that despite all this, there has been a general trend in the evolutionary process that can be of historical significance in the sense of producing implications that might guide analyses of history. Very briefly, there is no indication on a paleontological level of any major trends for the evolutionary process as a whole for the last billion or so years. Individual evolutionary lines may show an increase in, say, body size for several millions of years but then may decrease in size. A line may develop a more complex water- or waste-eliminating system but may then reduce this system drastically. The capacity for biochemical syntheses is very high in free-living protozoans but considerably lower in mammals or in parasitic protozoans. There was a history of advancing cephalization in most of the crustaceans but the barnacles have reversed this trend strongly. In short, most evolutionary lines have not maintained consistent directions of evolutionary change. For most postulated consistent directions of change, there is at least one evolutionary line that changes in an opposite sense.

Certainly, increased control over the environment and increased neural capacity are not general evolutionary trends. Witness the obvious fact that most living organisms survive beautifully with minimal intellectual accomplishments. The fact that man is better at intellectual operations might or might not appear impressive to a ciliate protozoan that can perform biochemical syntheses that make it free of the need for all but one vitamin, for example.

There is a persistence of the idea that complexity has increased during the course of evolution. The meaning of this assertion is not very clear. Morphological, biochemical, behavioral, neurological, and physiological complexity are each separately definable and may lead to contradictory assertions. The geometric complexity of the genital apparatus of free-living flatworms is enormous (Hyman, 1951), as is the capacity of Hawaiian fruitflies to produce new species (Carson et al., 1970).

Xenophanes said, "The Ethiopians make their gods black and snubnosed; the Thracians say theirs have blue eyes and red hair. . . . if oxen and horses or lions had hands, and could paint with their hands, and produce works of art as men do, horses would paint the forms of the gods like horses, and oxen like oxen, and make their bodies in the image of their several kinds," thereby pointing out the dangers of viewing the world from our own preconception of our own importance (Warner, 1950). There is no more evidence that man has a focal or teleological role in evolutionary history than that man's planet Earth is in a geographically central location in the cosmos. Ayala (1974) has discussed this question and notes the impossibility of assigning any internal, biological criterion of progress to evolution—although, once a decision has been made giving us a criterion of what we will consider progressive, we can ask whether or not individual evolutionary lines or the "average" of some assemblage of evolutionary lines has been progressive.

If we were to find some criterion of progress that would permit us to assert that all or even most evolutionary lines have been progressive, we could then ask if there is some mechanism responsible for this progress, and if we found such a mechanism, we could assert the existence of a trend. To my knowledge, no such criterion has been found.

There is some room for question in this area, however, since evolutionary events do share with historical events the property that certain changes have definite prerequisites. It is impossible to modify a wing until a wing has evolved, just as it is impossible to depose a monarch in the absence of a monarchy. It is possible to construct one kind of argument for a trend toward complexity on this basis, but I don't feel it to be a fruitful activity.

Notice that the fact that the present properties of organisms strongly qualify what will or will not be an advantageous change under particular environmental conditions does not impose a long term trend, since the environmental conditions are in no way constrained. In principle, any specialization or complexity of an organism can evolve into something more general or more simple under suitable selection—although it will almost certainly not return to an ancestral condition. We will see, however, that certain adaptations would require very peculiar and unlikely environmental circumstances for their complete reversal or elimination. In one sense, such adaptations act as valves or gates in evolutionary history. Once organisms have passed through them, their relation to the world is so changed that environmental forces act on them in a very different way than before. The capacity for terrestrial breathing is one example, and we will suggest that the development of human self-consciousness is another.

IV. EVOLUTION CONSIDERED AS AN EXISTENTIAL GAME

The fact that trends must generally be absent in the evolutionary process becomes apparent when one considers an analogy with game playing. For all possible universal evolutionary trends that have been suggested so far, biological counterexamples have been found. If, however, there were a trend, so that something was universally maximized in the evolutionary process, it would be possible to build a game theoretical model in which the organisms would be the players against nature and in which they should be acting as if the maximization of whatever it is that was involved in the trend is the payoff (Slobodkin, 1975; Slobodkin and Rapoport, 1974; Rapoport, 1975). But one peculiar property of most games tells us why this cannot be the case.

A payoff in a game is something that has value someplace other than in the place where the game is being played (except in a rather peculiar case;

see Section V). Money won at a dice game is primarily of value because it can purchase things someplace away from the crap table. It may, in addition, serve as a counter or a chip in the game, but for that purpose its place can be taken by poker chips or matchsticks. Its real meaning as a payoff is its purchasing power away from the game. Similar remarks can be made for almost any games and even for the rewards in athletic events (Huizinga, 1950).

The implication here is that if organisms were maximizing something as a universal evolutionary trend, they would be acting as if they had some place to spend their winnings, and, as a metaphysical point, they do not have any such place. That is, there is no way for evolutionary units to withdraw from the evolutionary game to spend what they have won. That may seem like simply a sophistic trick, but let us look at the implications of the fact that organisms cannot cash in their chips, staying within the broad context of game theory. Consider what it means about how organisms do in fact behave.

Making a fairly long story terribly short, there exists a class of existential games that corresponds in number, essentially, to the class of payoff games. The existential games are defined as those in which the payoff must be used in the game itself and cannot be cashed in. The way such a game is optimally played turns out to correspond very closely with what evolutionary units, in fact, seem to do. The optimal strategy in an existential game in which chips cannot be cashed is to minimize the stakes you play at each hand. That is, if you cannot win, what you do is try to persist—because you can lose. Extinction is equivalent to losing completely. Trying to persist means to risk as little as possible. Organisms seem to be playing a minimal-risk strategy. This will turn out to have major implications for the relation between history and evolution.

What actually happens in an evolutionary unit when the environment shifts and some disturbance occurs is that there are a series of simultaneously initiated responses, each one of which can, to some degree, deal with the kind of problem that the environment has presented. Unless the disturbance is completely novel, a series of behavioral and physiological responses are initiated. The elements in the series may be characterized by the relative rates at which they are activated and the degree to which they represent a major commitment of the organism's resources, in the sense of altering its capacity to deal with other environmental problems that are likely to arise.

If we rank the responses in terms of how much of a commitment they represent, we will find that those that represent the greater commitment take a longer time to be fully activated. This circumstance provides an opportunity for those response procedures that represent a minimal commitment to deal with the environmental problem. To the degree that minimal fast responses succeed in dealing with the problems appropriately, the full activation of the major commitments need not, and does not, occur.

The quickest way to get a sense of this phenomenon is to consider a pedagogic device that I have used: throwing chalk at a student during a lecture. The responses, depending on the age, sex, and character of the student, may be summarized. There is an eye blink and a sudden head movement, and these as a rule solve the immediate problem of flying chalk. Despite the fact that the problem has passed, the students typically show signs of some anxiety and an alertness in attitude and may show signs of circulatory changes. As long as 30 minutes after the chalk throwing, the student has a heart rate, an adrenal level, and a vasoconstriction pattern that is somewhat modified and an attention set that is quite distinct from that of a student who has not been the target of thrown chalk. The problem, however, was solved before most of the mechanisms listed really got started. The major responses that might have been needed don't occur. The student's healing capacity is not called on. The student typically is not angry enough either to run away or to strike me, and no one has ever dropped a course because of it. In short, a minor disturbance has been handled by a minor response. A lot of backup machinery was activated but did not really have time, between the occurrence of the disturbance and the solution of the problem caused by the disturbance, to be fully activated, or to alter the life of the organism in any serious fashion. It might be noted that the target students show some signs of persistent anxiety about my future behavior, but I will ignore this problem for the moment.

If it had been a bullet instead of chalk, the deeper responses would really have been of major importance, and I would hardly have needed to mention the response of blinking.

To the degree that an individual's response mechanisms fail, the individual either dies or fails to reproduce, and this result would quite probably constitute a change in the evolutionary characteristics of the population from which the individual was drawn. If, for example, the people who taught classes were in the custom of shooting at their students, and if there was some genetic relation between going to school and not going to school, we would diminish, in the population as a whole, the genetic class of people who go to schools where instructors shoot their students.

The whole structure of the anatomy and the physiology and the behavior of organisms tends to stand between the genes and their environment and tends to act as if it has the function of protecting the genetic properties of the organisms from the necessity of changing. To put it another way, the anatomy, physiology, and behavior of organisms can be thought of as the genes' way of avoiding having to change, so that we have a more-or-less universal set of properties of organisms in evolution that will mediate the path that evolutionary events take and will give us certain choice criteria for telling what an organism is likely to do or what a population of organisms is likely to do under a given perturbation without setting any definitive trend other than a minimization of the disruption of the organisms' existing genetic structure by the environmental events.

In those evolutionary lines in which learning capacity has developed, however, the story becomes more complicated, and this complication is discussed in Section V.

We might note here that a certain kind of predictability has become available. We can say that those environmental events that occur very rapidly and are highly transitory are optimally responded to by the development of a behavioral response mechanism, while those events that take somewhat longer time periods or represent deeper environmental changes should be responded to by deeper mechanisms on the part of the organism, so that we have defined, in a sense the class of evolutionary solutions that should transpire, given any particular environmental event.

V. THE PECULIAR EVOLUTIONARY STRATEGY OF MAN

I am going to contend that man and at least some other animals have a biological property of having an image of themselves to which they respond in the way that flatworms or protozoans respond to environmental perturbations. That is, in addition to the normal family of enviromental responses that we so briefly described, people have a set of internal properties that dictate their responses. Obviously, all animals have this to some extent. A hungry protozoan behaves differently than a well-fed protozoan. What I am refering to here is an elaboration of that property so that the sensory machinery itself is looking inward.

Conscious, normative, introspective self-images are a property of men set by their biology and this property seriously alters their response to evolutionary problems (Slobodkin, 1977).

When we say "set by their biology," we must be very clear. The tendency or propensity to develop a sense of self-awareness is biologically set. The content of the images that a man carries under the heading of self-awareness are almost certainly not biologically determined but are in fact determined by his cultural milieu.

It may be a trifle stronger than that. It may be the case that not only is there a propensity to maintain a self-image but also that if circumstances make it impossible to develop an internally consistent self-image, behavioral pathology may result.

The evolutionary background of this peculiarity of man seems to extend well back into the social mammals. Arboreal social mammals rely heavily on learned behavior to move through their habitat in an effective way. The fact of learning implies that the organism responds to present disturbances in terms of its own past history of the outcome of responding to disturbances. That is, as the efficacy of learning and the level of intelligence of animals increases, they are more and more responding to an intellectual construct of the world and more and more accepting information from the world as signals rather than perturbations. This progression reaches its epitome in man and some of the higher apes, particularly chimpanzees, orangutans, and gorillas, in which the intellectual construct of the world is at such a state of development that the animal can visualize itself as an object (Gallup, 1969, 1974).

I am not simply speculating about what goes on in the mind of a chimpanzee. I refer to specific experiments in which chimpanzees and orangutans are shown mirrors and respond to the mirror in the same way that a man responds to a mirror; that is, they groom themselves and adjust their own appearance in the mirror rather than treating the mirror image as another animal or as a competitor or as simply a picture of an animal. The apparent process by which this development of self-image may have evolved does not seem to require any departures from normal evolutionary arguments (Slobodkin, 1977).

What is of immediate interest is that once an animal sees itself as an object, its own ideas or fantasies about itself enter into the complex of forces that act on it in an evolutionary way. The self-image of men, chimps, and orangutans has a normative component, which is in fact how we can tell that they have a self-image. That is, a chimpanzee will attempt to wipe off color that has been placed on its skin, using a mirror. The color "doesn't belong there." It does not correspond to the chimpanzee's own image of what it ought to be, or if you object to that wording, simply say it does not correspond to the chimp's own image of itself. But once we say "its own image of itself," there is an implication of the normative.

Obviously, I cannot attempt to summarize the full range of speculation and theory about the nature of self. For our purpose, it may be sufficient to think of self in a behavioristic way, ignoring the neuroanatomical—and, to a degree, the philosophical—questions involved. Even the behavior will be dealt with in a highly cursory fashion, derived from the work of Mead (1934), Cooley (1964), and Merleau-Ponty (1964), among others.

Basically the self as I am considering it is not manifest in a newborn infant and develops only as the child communicates with other individuals. Initially in a very young infant, there are sensory responses and perhaps responses to signals of various kinds—laughing, crying, smiling. This period is followed by a period of imitation of sounds and of movements, which precedes the learning of the first words.

The behavioral manifestation of the self only occurs at around 2 or 3 years of age, when the child shows objective evidence that he can "see" himself as distinct from his physical environment, as distinct from other people—and as an object to which other people might respond. That is, Mead noted that a baby at the age of 2 or 3 will play with a companion and then, when the game is terminated, may continue to play alone and address himself by name or may take the role of the playmate with respect to himself. When a baby can take the role of another individual and address himself by name in that role, it is behaviorally apparent that he has an objective self-awareness of some kind. It has been suggested that the first concept of self-awareness arises from the baby's thinking of himself in the third person as a kind of "me" and then later identifying that third person with an "I." In order for him to do this, there must at some stage be a sense of a generalized "other." That is, self-image arises in a strongly social context.

The details of the development of the self are the subject of massive ongoing research in sociology, anthropology, psychology, and neurobiology. It is not my intention to review this research here. For the immediate purpose, it is sufficient that there is behavioral evidence of a self in humans and that humans can take the role of another and refer to themselves. That is, the assertion by a baby named Johnny that "Johnny is bad" is evidence of self-awareness. It must be emphasized that there is no evidence at all that self-awareness of a normal kind would arise in isolation. Also Gallup (1974) notes that in certain neurological disorders, patients cannot identify their own image in a mirror. Self-awareness develops out of the experience of communication with other individuals.

It is also clear that the idea of the self that is to be protected and cherished extends beyond the physical body and includes toys, objects, places, and persons who are in some sense incorporated into the self.

A man can decide that he is a worrier, he can decide that he is a saint, he can decide the normative state to which he aspires. Now, the concept of "can decide" seems to embroil us in various problems of free will and so forth. We need not enter into that area of controversy. All that is required is that men have a propensity, determined genetically, to develop a self-image. The "content" of that self-image need not be genetically determined in any sense. The self-image, once developed, acts as a normative standard. Precisely how the content of the self-image develops we leave open. On a common-sense basis, one would think that it would be a matter of details of upbringing. That is, you can raise a child in a context in which he develops for himself a self-image of high aggressiveness, or of high docility, or of any one of a variety of possible characteristics. Once this self-image has been developed, it becomes a major force in his behavior, and to the degree that the public representation of these images in speech and description are common to a culture or a time, they can become major historical and political forces.

I am making the very strong statement, at this point, that the biological nature of man is to have fantasies and to have images and to have dreams. There is no evidence that the content of these fantasies, images, and dreams is biologically determined by genetic differences between individuals or between groups of humans or that there is any fantasy that is biologically impossible to mankind as a whole.

This assertion is in contradistinction to the various statements of the innate aggressiveness of man, the innate cooperativeness of man, the innate territoriality of man, and so forth. That is, all of the assumed innate behavioral propensities of man, aside from those that are immediately tied to biological functions (e.g., the innate propensity to yawn when sleepy), seem to derive from the fantasies of their authors and seem to be descriptive, at their best, of the publicly shared fantasies of certain cultures. That is, the fact that the north Europeans may have prided themselves for centuries on their warriors and on their toughness and on their ability to drink and be generally rude says nothing one way or the other about the biological propensities of the northern Europeans compared to, say, Bushmen. The northern Europeans have been sharing a fantasy that men ought to be tough in certain ways, while the Bushmen have been sharing a different fantasy.

The fantasy takes on the role of at least a partial molder of history. The ability to fantasize in this way comes out of the evolutionary process, but the content of the fantasy does not derive directly from evolutionary laws. This is a strong assertion. I am saying that poetry is historically powerful. I am saying that the common-sense feeling that major historical events are initiated by ideas as well as by economic and perhaps ecological events is perfectly consonant with the biological facts and that one cannot, in particular, consider the most major historical events to be pushed or to have been directed by biological laws, or by biological events, but rather by the intellectual activities of men in their most private and poetic context.

This is not a new discovery, obviously. The Soviet Union has been imprisoning poets for years because of an awareness of this kind of thing and Plato made it quite clear that the forms of music and the forms of art are absolutely critical to the functioning of the state. All I have added here is the clarification of how this kind of fantasy relates to the biological nature of man, and in particular, I have denied the spate of recent speculations by Ardrey and Lorenz and Morris and so forth, which would assert that the content of man's self-image is itself biologically determined. There is absolutely no evidence that the content of a man's self-image is determined by his biology. We can just manage to say that the necessity to have a self-image is biologically determined.

What does this discussion of self-image imply for history and evolution? It says that it is impossible to assign major historical events to causes resident in biological evolutionary laws. That is, we cannot deduce the laws of history, if they exist, from the laws of evolution. This does not deny the possibility of constructing similes between historical events and evolutionary events or "laws." These have empirical meaning only in a most attenuated sense. For example, just as the evolutionary modification of wings is possible only in winged animals, the historical modification of wheels is possible only after the invention and adoption of the wheel. This fact can lead to such charming parallels between evolutionary and historical change as those noted in the history of military uniforms by Koenig (1975), in which he shows striking resemblances to the development of vestigial characters, for example, purely ornamental, nonfunctional buttonholes.

It has recently been suggested by Alexander (1975) and Wilson (1975) that while there may not be a biologically determined conscious self-image, there may be a deep biological determinism for humans to act in such a way as to maximize the perpetuation of genetic material identical with their own by providing aid to genealogical relatives who share this genetic material. Sahlins (1976) has provided many examples in which cultural definition of kinship and of behavior with reference to kin is such as to deny any validity to this concept. Neither Alexander nor Wilson nor Sahlins emphasizes the role of self-image as developed here.

Reviewing, we have shown that the evolutionary process is momentum-free and is driven by external perturbations, or external events, to which organisms by and large respond in a minimal fashion. People and some of the higher apes have interposed between themselves, their response machinery, and the environment a rather complex set of mental images and will respond to environmental disturbances to minimize disturbance to their mental image of the world as well as to minimize disturbances of a physiological kind. This tendency is probably so pronounced that the fantasies about mankind that men have are themselves a major evolutionary force. Certainly, the behavioral propensities of man derive much more from his ideas about what man is about than they do from any conceivable innate tendency to behave in certain modes.

Notice, we must distinguish between the biological behavioral responses and those responses related to such things as motivation. The biological responses we share with the bulk of animals. We blink, we eat, we sweat, and so on. The more interesting behavior about which some controversy can exist is behavior involving values, motives, goals—all those things that are embodied in man's image of himself. The only reasonable biological position at the moment is that that image is completely free to cultural determination and is not in any sense biologically given.

Therefore, the historian is unable to rely on biology to help him look at human historical trends. That is, once the self-image and communication about images and fantasies were fully established (in at least the last half million years), intellectual constructs became more important than evolution in determining history. On the other hand, however, the consequences of human dreams and human motives can have major evolutionary effects. We will see that some moral problems are exacerbated rather than solved by evolutionary laws. To put it another way, we cannot abrogate our personal responsibilities and get around problems by saying, "It is the nature of man." We are, however, confronted with biological consequences of human activity that set up difficulties for us.

VI. RESIDUAL EVOLUTIONARY PROBLEMS

In general, evolution is conservative. A gene-frequency change is a response to an enviromental change, and there are certain dangers to future survival associated with any change severe enough to result in gene-frequency changes. Certain human activities cause gene-frequency changes in human populations, and administrative decisions of one sort or another have to be made in conjunction with these changes. For example, imagine a human society in which there is always high infant mortality, always war, always misery, but the society is maintained within an ecological steady state and the war and the misery and the high infant mortality have been going on for a very long time. Then, if one avoids the death of children, if one eliminates war, if one alters the general state of misery, it is almost certain that one will cause gene-frequency changes.

That is, if you are going to be on the side of evolution, you will be on the side of the *status quo* almost always, and to make a change in the life pattern of any group of people will result in evolutionary shifts and will be (in a sense) counter to evolution. At that point, it matters very much whether you believe that evolution is a good that has to be maintained. If you believe this, it is part of your fantasy world; it does not come from the evolutionary process itself. That is, the future evolution of man will grow out of the fantasy world of man to some degree.

Gene-frequency change is the response to an adaptive failure and in that sense is to be avoided under most circumstances. This is not equivalent to saying that one deliberately avoids evolutionary change; it rather says that biological solutions can be very unpleasant indeed. Let's take a modern case in which biological evolution of man is now occurring. There is a famous genetic abnormality called the sickle-cell gene that produces abnormal hemoglobin. Individuals who are hybrid for this gene are immune to certain kinds of malaria. A double dose of this gene results in death from anemia. If malaria is all over the place, the majority of people that grow up to reproduce will be possessed of a single sickle-cell gene and will have a single normal gene, and they will have to reconcile themselves to one quarter of their children dying of anemia and almost one quarter dying of malaria but half their children dying of things other than anemia and malaria, that is, living a much longer life. We would expect that a few individuals not carrying the sickle-cell gene would survive the malaria, but these individuals may have to watch the overwhelming majority of their children die unless they marry someone carrying a sickle-cell gene, in which case only half of their children will be expected to die of malaria. All parents are therefore reconciled to the death of approximately half of their children. Notice that making immunity to malaria contingent on a gene that has these nasty properties of producing anemia is not an ideal solution, but this is the way it happens to have happened by evolution.

If the malarial infection is wiped out or the people are moved to some area where malaria is absent, all of the children of a non-sickle-celled parent will survive, but the survival of the children of the sickle-cell parent will vary with whether or not the other parent carries the sickle-cell gene. Two social attitudes are now possible: one relies on the evolutionary process to deal with the social problem and in that sense is a natural process, and the other is built on a human ethical image and on a rejection of the evolutionary solution.

Those who adopt the position of letting evolution solve the problem correctly observe that since carriers of sickle-cell genes will be lower in reproductive capacity than other members of the population, each generation will have a smaller and smaller proportion of such individuals, so that the problem will eventually become a very minor one. When only 1% of the individuals have the sickle-cell gene, only 1 marriage out of 10,000 will give rise to children that die from sickle-cell disease, if marriage partners are chosen at random.

An alternative approach is to develop a diagnostic test for the presence of sickle-cell genes so that a person carrying the gene can be warned and then left with three personal choices. The first is to ignore the problem. The second is to marry a person known to be free of the sickle-cell gene; this solution has the advantage that no children will die of sickle-cell anemia but has the disadvantage that half the children will be carriers of the sickle-cell gene. The third choice is to marry a person with sickle-cell genes and avoid reproducing. Under this system, sickle-cell genes will be expected to decline in relative frequency in the population. The social implication of telling people that they carry the sickle-cell gene is that the frequency of the gene will be slightly reduced in the population in the course of time and that in the meanwhile, those people who suffer from either sickle-cell anemia or the sociological and psychological consequences of knowing that they are sickle-cell-gene carriers will just have to accept their role in evolutionary history.

A third general alternative is to consider that sickle-cell anemia is a hereditary disease like any of the other known biochemical-defect diseases and then attempt to cure the sickle-cell-anemia condition. That is, one sets up a medical program that, if successful, will transform the sickle-cell-anemia condition into one, like diabetes, that an individual can deal with in normal life by taking medication. If such a program were successful, it would give sickle-cell-trait parents the same reproductive potential as normal-hemoglobin parents, other things being equal, and might also permit individuals with a double dose of sickle-cell gene to reproduce. This approach would maintain the sickle-cell gene in its present ratio in the population but would also eliminate a great deal of personal misery.

At that point, the issue is joined. Either you have to think of man as an evolving organism subject to the laws of evolution—and those who for one reason or another must suffer personally for this simply have to suffer because that is nature's way—or you make the clean split that says that man has his own image of the world and that from that image of the world he builds his own environment as far as possible. It is within that environment that we define the notion of fitness and of selective value. We build our definition in terms of the moral values that we have adopted out of the freedom given us by evolution to make such decisions.

The suffering of a parent or the suffering of a baby due to the sickle-cell condition can now be thought of either as a consequence of evolutionary law or as an inadequacy in human medicine and human society. I think my preference is obvious. In fact, I believe that what I have done is to reaffirm on the basis of the facts of evolution the curious metaphysical bind of men in which we are condemned to freedom. We are condemned to deal with moral and aesthetic issues as moral and aesthetic issues and cannot relegate our naturally given responsibility to some kind of natural law.

VII. ACKNOWLEDGMENTS

This paper was written while the author was supported as a Guggenheim Fellow. Research relevant to this paper has been supported by the

National Science Foundation (NSF 31-756A). It is contribution No. 185 of the Ecology and Evolution Program, State University of New York, Stony Brook. It is based on lectures given to the American Historical Society at their 1974 meeting and at the Hastings Institute, Hastings, New York.

VIII. REFERENCES

Alexander, R. D. (1974). The evolution of social behavior. *Annu. Rev. Ecol. Syst.* **5**:325–383.
Alexander, R. D. (1975). The search for a general theory of behavior. *Behav. Sci.* **20**(2):77–100.
Ardrey, R. (1966). *The Territorial Imperative*. Atheneum, New York.
Ayala, F. J. (1974). The concept of biological progress. In Ayala, F. J., and Dobzhansky, T. (eds.), *Studies in the Philosophy of Biology*, Macmillan, London.
Barzun, J. (1958). *Darwin, Marx, and Wagner: A Critique of a Heritage*, Doubleday, New York.
Braudel, F. (1972). *The Mediterranean World in the Age of Philip II*, translated from the French by S. Reynolds, Harper and Row, New York.
Carson, H., Hardy, D. E., Speith, H. T., and Stone, W. S. (1970). The evolutionary biology of the Hawaiian Drosophilidae. In Hecht, M. K., and Steere, W. C. (eds.), *Essays in Evolution and Genetics in Honor of Theodosius Dobzhansky*. Appleton-Century-Crofts, New York, pp. 437–543.
Cervantes Saavedra, Miguel de. (1968). El retablo de las maravillas, *Entremeses*, Prologo y notas de J. A. Franch, Las America Publishing Co., New York.
Cooley, C. H. (1964). *Human Nature and the Social Order*, Shocken Books, New York.
Fisher, R. A. (1930). *The Genetical Theory of Natural Selection*, Clarendon, Oxford.
Gallup, Jr., G. G. (1969). Chimpanzees: Self-recognition. *Science* **167**:86–87.
Gallup, Jr., G. G. (1974). Toward an operational definition of self-awareness. In Tuttle, R. H. (ed.), *IXth International Congress of Anthropological and Ethnological Sciences*, Primatology Session, Subsection D, Mouton Press, The Hague.
Grant, M. (1916). *The Passing of the Great Race or the Racial Basis of European History*, Scribner's, New York.
Huizinga, J. (1950). *Homo Ludens: A Study of the Play Element in Culture*, Beacon Press, Boston.
Hyman, L. H. (1951). *The Invertebrates: Platyhelminthes and Rhynchocoela, Vol. 2, The Acoelomate Bilateria;* McGraw-Hill, New York.
Kamin, L. J. (1974). *The Science and Politics of I.Q.*, Halsted Press Division, Wiley, New York.
Koenig, O. (1975). Biologie der uniform. In Ditfurth, H. v. (ed.), *Evolution ein Querschnitt der Forschung*, Hoffman und Campe, Hamburg, pp. 175–211.
Lewontin, R. C. (1974). *The Genetic Basis of Evolutionary Change*. Columbia University Press, New York.
Mead, G. H. (1934). *Mind, Self and Society from the Standpoint of a Social Behaviorist*, Morris, C. W. (ed.), University of Chicago Press, Chicago.
Merleau-Ponty, J. (1964). *The Primacy of Perception*, Northwestern University Press, Evanston, Ill.
Morris, D. (ed.) 1967. *Primate Ethology: Essays on the Sociosexual Behavior of Apes and Monkeys*, Aldine, Chicago.

Rapoport, A. (ed.). (1975). Uses of game theoretic models in biology. *General Systems: Yearbook of the Society for General Systems Research,* Volume 20, Society for General Systems Research, Washington, D.C. pp. 49–58.

Roth, C. (1964). *The Spanish Inquisition,* Norton, New York.

Sahlins, M. (1976). *The Use and Abuse of Biology, an Anthropological Critique of Sociobiology,* University of Michigan Press, Ann Arbor, Mich.

Slobodkin, L. B. (1961). *Growth and Regulation of Animal Populations,* Holt, Rinehart, and Winston, New York.

Slobodkin, L. B. (1968). Toward a predictive theory of evolution. In Lewontin, R. (ed.), *Population Biology and Evolution,* Syracuse University Press, Syracuse, N.Y., pp. 187–205.

Slobodkin, L. B. (1975). Die Strategie der Evolution. In Ditfurth, H. v. (ed.), *Evolution: Ein Querschnitt der Forschung.* Hoffman und Campe, Hamburg, pp. 33–55.

Slobodkin, L. B. (1978). The peculiar evolutionary strategy of man. *Transactions of the Boston Colloquium of Philosophy of Science,* Vol. 31, in press.

Slobodkin, L. B., and Rapoport, A. (1974). An optimal strategy of evolution. *Q. Rev. Biol.* **49**(3):181–200.

Spengler, O. (1934). *The Decline of the West,* Knopf, New York.

Trivers, R. L. (1971). The evolution of reciprocal altruism. *Q. Rev. Biol.* **46**(4):35–57.

Vayda, A. (1976). *War in Ecological Perspective,* Plenum, New York.

Waddington, C. H. (1957). *The Strategy of the Genes,* Allen and Unwin, London.

Warner, R. (1950). *The Greek Philosophers,* Menton, New York.

Wilson, E. O. (1975). *Sociobiology: The New Synthesis.* Belknap Press of Harvard University Press, Cambridge, Mass.

INDEX